HUNDRED Million Dollar
BITCOIN

Beginners' Course

SuPiArO
Crypto Starter Pack

Note

ISBN: 9798388418333
Independently published on Amazon.

This course is created with the help of artificial intelligence and is targeted at high school age learners and everyone else interested in understanding the basics of how Bitcoin works and its importance in today's world.

Copyright © 2023 SuPiArO
All rights reserved.

You're free to use this book's content for academic purposes only. No profitable copying or distribution is permitted without the author's consent.
Contact: +2347068435789

About this Course

Welcome to the Hundred Million Dollar Bitcoin for Beginners course! This course is designed for high school age students who are interested in learning about the basics of Bitcoin, the popular digital cryptocurrency that has gained worldwide recognition over the past decade.

The purpose of this course is to provide a comprehensive overview of Bitcoin, its technology, and its monetary value. By the end of this course, you will have a solid understanding of how Bitcoin works, how it is mined, and how it is used. You will also learn about the security measures that are in place to protect the Bitcoin network and how to safely store and trade Bitcoin by yourself.

Why is learning about Bitcoin important?

As we move towards a more digital future, understanding digital currencies like Bitcoin will become increasingly relevant. Bitcoin is already being used as a form of payment by some businesses and individuals, and its use is likely to continue to grow in the coming years. Additionally, Bitcoin's decentralized and transparent nature has the potential to disrupt traditional financial systems and bring financial freedom to people who may not have had access to it before.

The course is divided into five modules:

Module A, Introduction to Bitcoin, covers the basics of what Bitcoin is, how it works, and its key features.

In Module B, The Technology behind Bitcoin, you will learn about the blockchain, mining, and nodes.

Module C, Bitcoin Network's Security, covers the security measures that are in place to protect the Bitcoin network, including the 51% attack, double spending, and privacy concerns.

Module D, Bitcoin's Monetary Value, covers Bitcoin as a form of money and its implications as a store of value.

Finally, in Module E, Holding and Trading Bitcoin, you will learn how to buy, store, and trade Bitcoin, as well as the risks associated with doing so.

Each lesson in this course is designed to be interactive and engaging, with real-world examples, where necessary. By the end of the course, you will have a solid understanding of Bitcoin basics and the potential it holds for our digital future.

Whether you're interested in investing in Bitcoin, using it as a form of payment, or simply curious about the technology behind it, this course is for you.

Join me on this exciting journey into the world of Bitcoin!

Happy Learning.

Table of Content

Contents

Note 1
About this Course 2
Table of Content 5
Module A 9
Introduction to Bitcoin 9
What is Bitcoin? 10
Lesson 2 13
How does Bitcoin work? 13
Lesson 3 17
Bitcoin's Origin 17
Lesson 4 21
Bitcoin versus Fiat 21
Lesson 5 27
Bitcoin's Key Features 27
Lesson 6 33
Bitcoin's Use Cases 33
Module B 39
The Technology behind Bitcoin 39
Lesson 7 41
The Bitcoin Blockchain 41
Lesson 8 45

Bitcoin Mining 45

Lesson 9 49

Bitcoin Nodes 49

Lesson 10 53

Bitcoin Halving 53

Lesson 11 57

How Bitcoin Transactions Work 57

Lesson 12 63

UTXOs and Change Addresses 63

Lesson 13 69

Bitcoin Forks 69

BIPs 72

Module C 77

Bitcoin Network Security 77

Lesson 15 78

Bitcoin's Security 78

Lesson 16 84

51% Attack 84

Lesson 17 87

Double Spending in Bitcoin 87

Lesson 18 91

Privacy in Bitcoin 91

Lesson 19 96

Decentralization in Bitcoin 96

Lesson 20 99

Why Bitcoin Needs so much Energy 99
Module D 104
Bitcoin's Monetary Value 104
Lesson 21 106
Bitcoin as Money 106
Lesson 22 113
Bitcoin as Legal Tender 113
Lesson 23 117
Bitcoin Tokenomics 117
Lesson 24 121
Bitcoin as Store of Value 121
Lesson 25 127
Implications of Bitcoin Scarcity 127
Lesson 26 131
Bitcoin Market Dominance 131
Module E 135
Holding and Trading Bitcoin 135
Lesson 27 137
How to Buy Bitcoin 137
Lesson 28 143
Storing Bitcoin Securely 143
Lesson 29 149
Sending and Receiving Bitcoin 149
Lesson 30 155
Trading Bitcoin 155

Lesson 31 162
Risks to watch 162
Course Endnote 171

Module A
Introduction to Bitcoin

In this module, I will introduce you to the revolutionary digital currency that is changing the way we think about money and finance.

Throughout this module, you will learn about the basics of Bitcoin, including its origins, key features, and various use cases. You will gain a fundamental understanding of what Bitcoin is, how it works, and how it compares to traditional fiat currencies.

We will explore the origins of Bitcoin, including the identity of its anonymous creator, and the historical events that led to its creation. You will also learn about the key features of Bitcoin that set it apart from traditional currencies, such as decentralization and transparency.

We will examine Bitcoin's various use cases, including its potential as a means of payment, investment, and store of value. You will also learn about the advantages and disadvantages of using Bitcoin compared to traditional fiat currencies.

By the end of this module, you will have a solid understanding of the fundamentals of Bitcoin, setting the foundation for your continued learning in the following modules.

Get ready to dive into the exciting world of Bitcoin!

Lesson 1
What is Bitcoin?

Bitcoin is a digital currency that was created as an alternative to fiat currencies, such as the US dollar, the British pound, or the Nigerian naira. It is a virtual currency, meaning that it only exists in the digital realm and cannot be physically held or seen. Unlike fiat currencies, which are controlled by central authorities, such as governments or central banks, Bitcoin is decentralized. This means that it is not controlled by any single entity. This makes it a unique form of currency that operates differently from traditional forms of money.

Bitcoin was the first cryptocurrency ever created, and it remains the most popular and valuable cryptocurrency in existence today. Cryptocurrencies, in general, are a form of digital money that use cryptography, a technique of encoding and decoding information, to provide security and other functionalities. There are currently thousands of different types of cryptocurrencies in circulation, each with its own unique features and uses.

Bitcoin can be used just like fiat currencies, to buy goods and services or to store value. However, unlike fiat currencies, there are no barriers to entry. Anyone with an internet connection can participate in the Bitcoin network and use it to send or receive money without seeking permission from any central authority, such as a bank or government.

Bitcoin operates on a technology called the blockchain, which is essentially a public record book where individual records of transactions are permanently stored. These records cannot be changed or deleted once they are entered into the blockchain, and anyone can add an entry to the record book by following a set of rules.

This makes Bitcoin a peer-to-peer means of making financial payments for goods or services, with zero support for interference by middlemen or third parties. This is similar to the way physical cash work, except that Bitcoin is digital. It is a decentralized way to send or receive value among individuals who do not need to trust each other. All payments are recorded in a transparent, secure, and public database. Anyone can access the information on this database, but no one's identity is revealed by the record, making Bitcoin pseudonymous and users' identity anonymous by choice.

Bitcoin is not controlled by any single person, government, or company. The users who participate in the Bitcoin network control it collectively. Every participant keeps a complete or partial record of the blockchain on their own computer or phone, and all participants around the world work together to coordinate the blockchain seamlessly. They must agree on the information to be added to the record book before it is permanently added.

Summary

Bitcoin is a peer-to-peer electronic payment system that is not controlled by any central authority aside from its users. It enables users to send and receive money in the form of Bitcoin without the need for trust or third-party permission. It operates on a decentralized technology called the blockchain, where individual records of transactions are permanently stored in a transparent, secure, and public database. Although it is a form of digital currency, it has some similarities to physical cash, such as the ability to exchange it between two people without the need for a bank. However, it is superior to physical cash in that no one can declare it worthless except its users, and no one can prevent you from using it.

Lesson 2
How does Bitcoin work?

Bitcoin relies on a decentralized network called the Bitcoin blockchain, which uses cryptography to secure transactions and create new units of currency. In this lesson, I'll explain how Bitcoin works as a means of payment.

Bitcoin vs bitcoin

Before we dive in, it's important to understand the difference between Bitcoin and bitcoin. When we say "Bitcoin" with a capital B, we're referring to the protocol that underlies the digital currency. When we say "bitcoin" with a lowercase b, we're referring to the individual units of currency that can be exchanged on the Bitcoin network. These individual units are often denoted as BTC.

How Bitcoin Transactions Work

Let's say you want to buy something using bitcoin. Here's how the transaction would work:

Step 1: Get a Bitcoin Wallet

The first thing you'll need is a Bitcoin wallet. This is a digital wallet that stores your bitcoin and allows you to send and receive payments. There are many different types of Bitcoin wallets, but they all work in essentially the same way.

Step 2: Find a Merchant That Accepts Bitcoin

Next, you'll need to find a merchant that accepts bitcoin. This could be an online store, a physical store, or an individual who's selling something.

Step 3: Send Bitcoin to the Merchant

Once you've found a merchant and selected the item you want to buy, you'll need to send them the correct amount of bitcoin. To do this, you'll need to enter the merchant's Bitcoin address (which is a long string of numbers and letters that uniquely identifies their Bitcoin wallet) into your own Bitcoin wallet. You'll also need to enter the amount of bitcoin you want to send and pay a small transaction fee to incentivize Bitcoin miners to process your transaction.

Step 4: Wait for Confirmation

After you've sent the bitcoin, you'll need to wait for the transaction to be confirmed by the Bitcoin network. This usually takes a few minutes, but it can take longer if the network is congested. Once the transaction is confirmed, the merchant will receive the bitcoin and can then ship your item.

Step 5: Verify the Transaction

To verify that the transaction was successful, you can check your Bitcoin wallet to see if the correct amount of bitcoin has been deducted. You can also check the Bitcoin blockchain, which is a public ledger that records all Bitcoin transactions. By looking up the transaction hash (a unique identifier for the transaction) on the blockchain, you can confirm that the transaction was processed and that the correct amount of bitcoin was sent to the merchant.

Benefits of Using Bitcoin for payment.

There are several benefits to using Bitcoin instead of traditional payment methods:

Decentralization - Bitcoin is not controlled by any central authority, which means that it's not subject to government or corporate influence.

Anonymity - While Bitcoin transactions are not completely anonymous, they do offer a higher level of privacy than traditional payment methods.

Lower transaction fees - Bitcoin transactions typically have lower fees than credit card transactions or wire transfers. This is especially important if you're sending a large amount of money.

Fast transactions - Bitcoin transactions are processed almost instantly, which means that you don't have to wait for days or weeks for your payment to go through.

Global reach - Bitcoin can be used to send and receive payments anywhere in the world, without the need for currency conversion. There are no barriers.

Summary

Bitcoin is a digital currency that uses a decentralized network called the Bitcoin blockchain to secure payment transactions. A Bitcoin wallet is required to send and receive payments, and transactions are verified by the Bitcoin network. Benefits of using Bitcoin include decentralization, anonymity, lower transaction fees, fast transactions, and global reach. It's important to understand the difference between Bitcoin and bitcoin (BTC), as Bitcoin refers to the protocol while bitcoin refers to the units of currency.

Lesson 3
Bitcoin's Origin

Where did Bitcoin come from, and who created it? In this lesson, we will explore the origins of Bitcoin and its creator, Satoshi Nakamoto.

The Beginning of Bitcoin

On August 18, 2008, an anonymous individual registered the domain name bitcoin.org, and about six weeks later, on October 31 of the same year, the famous bitcoin whitepaper began to circulate online. These two events mark the beginning of the existence of Bitcoin.

The author of the Bitcoin whitepaper is named Satoshi Nakamoto, but whether it is a single person or a group of individuals that go by that name, it is not certain. The collective pronoun "we" was frequently used in the whitepaper, suggesting that Satoshi Nakamoto may not be a single individual.

The Identity of Satoshi Nakamoto

The identity of Satoshi Nakamoto is unknown, and there have been speculations that attempt to associate the pseudonymous name with known individuals who appear to exhibit a close semblance to who Satoshi Nakamoto was.

For instance, Hal Finney was one of the earliest people to communicate with Satoshi Nakamoto and even received the first bitcoin transaction from Satoshi himself. Many people believe he was the real Satoshi Nakamoto, but he denied this and died in 2013.

There are several others who are associated with the bitcoin creator. A man named Dorian Satoshi Nakamoto, who happened to live few blocks away from where Hal Finney lived, is also believed by many to be the actual creator of bitcoin, especially given his background in physics.

Other names rumored to be Satoshi include Nick Szabo, Wei Dai, and Adam Back. Both Wei Dai and Adam Back were referenced in the Bitcoin whitepaper. Not to mention a man named Craig Wright who claims to have written the Bitcoin whitepaper himself and says he's Satoshi Nakamoto and may have claimed a patent to the bitcoin whitepaper.

The First Bitcoin Block

On January 3, 2009, the first bitcoin block was mined by Satoshi Nakamoto, and it was named the Genesis Block. It contains a timestamp message referencing a newspaper report about bailout for banks experiencing financial crises in the UK.

It took about six days for the next Bitcoin block to be mined, and the BITCOIN blockchain went public on January 9, 2009. It was the day Satoshi shared his invention publicly, and on January 11, Hal Finney tweeted, "Running Bitcoin."

Why Was Bitcoin Created?

Bitcoin was created to be a decentralized digital currency that would allow for secure and transparent transactions without the need for a central authority or third-party intermediary. Before the creation of Bitcoin, there had been attempts to create digital currencies, but these had all failed due to a lack of trust and the need for a central authority to oversee the transactions.

Bitcoin was also the first digital currency to solve the problem of "double spending." Double spending refers to when an individual spends a currency twice, which would be the case for digital Fiat currencies in your Bank app if the Banks get hacked in any way. People will be able to send the same currency to several people indefinitely. However, the use of blockchain makes this impossible with bitcoin.

Summary

Bitcoin was created by a person or group of persons that go by the name Satoshi Nakamoto, and they mined the first Bitcoin block on January 3, 2009. According to the Bitcoin whitepaper released months earlier, bitcoin exists to enable peer-to-peer electronic transfer of value without the need for trust in a central authority.

Lesson 4
Bitcoin versus Fiat

But how is Bitcoin different from Fiat currencies?

Fiat currencies are the paper notes and coins that are issued and backed by a government. Examples of Fiat currencies include the US dollar, British pound, Nigerian naira, and others. Unlike gold and other commodities, Fiat currencies have no inherent value. Instead, their value is derived from the trust people have in the government that issued them.

On the other hand, Bitcoin is a decentralized digital currency that operates on a peer-to-peer network. It is not issued or backed by any government or central authority. Instead, it is created and maintained by a network of users around the world. So what are the key differences between Bitcoin and Fiat currencies? Let's take a closer look.

Issuance and Distribution

One of the biggest differences between Bitcoin and Fiat currencies is how they are issued and distributed. Fiat currencies are issued and controlled by central governments, which means that governments have complete control over the money supply. This also means that governments can manipulate the money supply to control inflation, stimulate the economy, and more.

Bitcoin, on the other hand, is decentralized, which means that it is not controlled by any central authority. Instead, it is created through a process called mining, which involves solving complex mathematical problems. Bitcoin mining is open to anyone with the necessary hardware and software, and the rewards for mining are distributed to the network of users.

Supply

Another major difference between Bitcoin and Fiat currencies is the supply. Fiat currencies have an unlimited supply, as governments can print as much money as they need. This can lead to inflation, which can reduce the value of the currency over time.

Bitcoin, on the other hand, has a limited supply. There will only ever be 21 million Bitcoins in circulation, and this cap is hard-coded into the Bitcoin protocol. This means that Bitcoin is a deflationary currency, which means that its value will increase over time.

Purchasing Power

Another advantage of Bitcoin over Fiat currencies is its deflationary nature, which means that its purchasing power increases over time. With Fiat currencies, the opposite is true. The value of Fiat currencies decreases over time due to inflation. If you save $10,000 for ten years, the amount of what it can buy reduces because its purchasing power goes down. But what 1 Bitcoin can purchase increases over time as evident in the last decade of its existence.

Transferability

Bitcoin is also more easily transferable than Fiat currencies. It can be sent from one user to another anywhere in the world without the need for intermediaries like banks. This means that Bitcoin transactions can be completed faster and at lower costs than traditional Fiat transactions. In contrast, the transfer of Fiat currencies often requires the involvement of financial institutions, which can lead to delays and additional fees.

Accessibility

While not everyone has access to Fiat currencies, Bitcoin is accessible to anyone with an internet connection. This means that people in countries with unstable currencies or limited access to traditional banking systems can still use Bitcoin to transact with others around the world. Bitcoin is banking the unbanked population of the world!

Borderless

Another advantage of Bitcoin over Fiat currencies is its borderless nature. Fiat currencies are often only accepted in certain countries or regions, which can make cross-border transactions difficult. Bitcoin, on the other hand, is universal and can be used for cross-border payments. Although it is not yet a legal tender in most countries, anyone anywhere can accept it and use it for transactions.

Anonymity

Bitcoin transactions can be relatively anonymous, as users can transact using only a pseudonym. This means that there is no need for identity verification or disclosure of personal information. In contrast, the use of Fiat currencies is often accompanied by the requirement to provide personal identification. Financial institutions and payment processors usually require individuals to disclose their identity before they can transact with Fiat currencies. This means that users are not entirely anonymous when using fiat currencies. The identity verification process can be quite cumbersome, and it may not be possible to transact without exposing one's true identity to financial institutions or payment recipients. Additionally, financial institutions and governments may track and monitor Fiat transactions for various reasons, including taxation and anti-money laundering efforts.

Transparency

One of the biggest advantages of Bitcoin over fiat currencies is transparency. Bitcoin transactions are recorded on the Bitcoin Blockchain, which is a decentralized public ledger. This means that anyone can view and monitor transactions in real time. The transparent nature of Bitcoin transactions ensures that there is a high level of accountability, and it discourages fraud and corruption. On the other hand, Fiat transactions are not available to the public, and institutions may mishandle money entrusted to them without public knowledge.

Uncensorable

Bitcoin transactions are uncensorable, meaning that they cannot be stopped or reversed by any government or financial institution. Once a Bitcoin transaction has been recorded on the blockchain, it cannot be altered or deleted. This feature makes Bitcoin transactions more secure and reliable compared to Fiat transactions, which can be subject to censorship and manipulation by governments or financial institutions. Even if your government clamps down on freedom of citizens, they cannot prevent anyone from using Bitcoin as long as that person has internet connectivity.

Summary

Bitcoin is a decentralized digital currency that operates on a peer-to-peer network, while Fiat currencies are paper notes and coins issued and backed by a government. One of the biggest differences between Bitcoin and Fiat currencies is how they are issued and distributed. Fiat currencies are issued and controlled by central governments, while Bitcoin is created through a process called mining, which involves solving complex mathematical problems. Another major difference is the supply. Fiat currencies have an unlimited supply, while Bitcoin has a limited supply of only 21 million. Bitcoin is also more easily transferable and accessible than Fiat currencies, and it is borderless, making cross-border transactions easy. Bitcoin transactions can be relatively anonymous, and its transparency ensures a high level of accountability. Finally, Bitcoin transactions are uncensorable, making them more secure and reliable compared to Fiat transactions.

Lesson 5
Bitcoin's Key Features

Unlike traditional currencies, Bitcoin is decentralized, meaning that it is not controlled by any government, bank, or central authority. Instead, it is underpinned by a peer-to-peer network that allows users to transact directly with each other without the need for intermediaries. In this lesson, we will explore the key features that make Bitcoin unique and explain why it is so important for the future of finance.

Decentralization

One of the most significant features of Bitcoin is its decentralization. Because it is not controlled by any central authority, it is difficult for any single entity to manipulate the network or interfere with transactions. This makes it a truly democratic system that is open to all, regardless of their background or financial status.

Permissionless

Bitcoin is a permissionless system, which means that anyone can use it without needing permission from any authority. This makes it an inclusive financial system that is accessible to anyone with an internet connection. It doesn't matter where you are or who you are, you can use Bitcoin to transact with anyone in the world.

Trustless

Trust is not required to use Bitcoin because it is secured using cryptography. This means that users can transact directly with each other without the need to trust each other or a third party to facilitate the transaction. This is a significant departure from traditional financial systems, which require users to trust banks or other intermediaries to facilitate transactions.

Censorship Resistant

Because it is decentralized and trustless, it is difficult for any single entity to censor or block transactions on the Bitcoin network. This makes it a censorship-resistant platform that can be used to facilitate transactions without fear or favour. This is particularly important in countries where the government may attempt to control or censor financial transactions.

Scarcity

There is a limited supply of bitcoin that will ever be created. The total number of bitcoins that will be created is capped at 21 million, with approximately 19.1 million already in circulation as of January 2023. This makes bitcoin a scarce asset, which can drive demand for it and potentially contribute to its value. The limited supply of Bitcoin means that it is not subject to inflation, which is a significant problem for traditional fiat currencies.

Transparency

Bitcoin transactions are recorded on a public ledger called the blockchain. This means that all transactions are transparent and can be viewed by anyone. This is a significant departure from traditional financial systems, which are often opaque and difficult to understand.

Anonymity

While transactions on the Bitcoin network are transparent, the parties involved in the transactions are not publicly identified. This means that users can transact with each other anonymously, preserving their privacy. This is particularly important for people who may not want their financial transactions to be publicly visible.

Irreversibility

Once a Bitcoin transaction has been recorded on the blockchain, it cannot be reversed. This means that it is important for users to be careful when sending bitcoin, as there is no way to recover it if it is sent to the wrong address. However, this feature also makes Bitcoin more secure, as it makes it difficult for fraudsters to reverse legitimate transactions, preventing double spending.

Immutability

The blockchain is a secure and immutable record of all Bitcoin transactions. This means that it is difficult to alter or tamper with the transaction history, ensuring the integrity of the network. This is an important feature for people who may not trust traditional financial systems and want a more secure way to transact.

Divisibility

Bitcoin can be divided into smaller units called satoshi. One bitcoin is equal to 100 million satoshi, which allows for transactions to be made in very small amounts. This makes it a versatile currency that can be used for both small and large transactions.

Fungibility

One of the unique features of bitcoin is its fungibility, which means that each unit of bitcoin is interchangeable with another. This makes it a more versatile and useful asset for exchange since it is not tied to any specific asset or identity. Fungibility ensures that bitcoin can be exchanged without any loss of value, just like cash. 1 bitcoin is always 1 bitcoin.

Portability

Another advantage of bitcoin is its portability. Unlike cash, which can be cumbersome and risky to carry around in large amounts, bitcoin can be easily transferred and stored digitally. This makes it a highly portable asset that can be accessed from anywhere with an internet connection. It is possible to store bitcoin in a digital wallet, which can be accessed from a smart phone or computer.

Fast transactions

Bitcoin transactions are generally processed very quickly, with most transactions being confirmed within a few minutes. This makes it a fast and efficient way to transfer value, especially when compared to traditional financial services such as wire transfers or credit card transactions. Bitcoin transactions are processed by a decentralized network of computers around the world, which eliminates the need for intermediaries and speeds up the process.

Cheap transactions

Another advantage of bitcoin is that transactions can be made at a very low cost, especially when compared to traditional financial services such as wire transfers or credit card transactions. The elimination of intermediaries makes the system more efficient, which results in lower transaction fees. In addition, bitcoin transactions do not require physical infrastructure, which further reduces costs. Someone can send a billion dollar worth of bitcoin and only pay $1 in fees!

Summary

Bitcoin is a decentralized digital currency that offers several unique features that make it stand out from traditional financial systems. Bitcoin is designed to resist censorship and centralized control, making it a more secure and transparent financial system. It is easily interchangeable and divisible, making it a versatile asset for exchange. Bitcoin is also highly portable, making it accessible from anywhere with an internet connection. Bitcoin transactions are fast and efficient, with low transaction fees, thanks to the elimination of intermediaries. Finally, bitcoin is a scarce asset, with a limited supply of 21 million bitcoins that will ever be created, which potentially contributes to its value over time.

Lesson 6
Bitcoin's Use Cases

Bitcoin has been gaining popularity and utility over the years due to its decentralized nature. Here are some potential uses of Bitcoin:

Low Cost Transactions

Bitcoin is a low-cost alternative to traditional financial transactions because it operates on a decentralized network, which means that it is not subject to fees and regulations seen with traditional financial institutions. Thus, it can be used to carry out transactions without incurring exorbitant prices.

Cross-border Payments

Bitcoin has simplified international payments and trades because there is no need to worry about foreign exchange of one's local currencies, which often incur exorbitant costs and delays.

Private Transactions

Bitcoin transactions are recorded on a public ledger called the blockchain, but user identities are pseudonymous to the public, meaning that their real identities are anonymous. Thus, anyone who does not want to reveal their personal information in a transaction can ensure their privacy via peer-to-peer transfer on the Bitcoin network. Techniques such as currency mixing can also increase the privacy of Bitcoin transactions, though this may be illegal in some jurisdictions.

Boycott Censorship

Bitcoin is decentralized and not controlled by any government or institution, so no government can stop Bitcoin transactions. It can be used to bypass censorship of any sort, making it freedom money. This is important for Bitcoin users living in countries where freedom is subject to the whims of dictators. For instance, in Nigeria in 2020, rioters against police brutality were able to rally financial support from around the world using Bitcoin when the government froze their traditional bank accounts.

Financial Inclusion

Banking the Unbanked: Not everyone has access to the traditional financial systems, and banks are not available in most remote communities. However, with the rise in internet accessibility, people in the innermost communities in the world can use Bitcoin for transactions. Bitcoin also makes it possible for those who are prohibited from the traditional financial systems to gain access to financial services because Bitcoin does not have any entry barriers.

Anti-Corruption

While identities on the Bitcoin Blockchain are pseu-donymous, transaction records are publicly viewable. The movement of funds can be traced and tracked in real time by anyone. If public funds are put in Bitcoin, it will be impossible for corrupt individuals to move and hide funds secretly as is the case with fiat, especially with cash. Even the movement of stolen funds can be monitored until it leads to a centralized exchange where the latest receiver can be used to track the real identity of the fraudsters.

Payment for Goods and Services

Bitcoin is a digital currency that can be used to make payments both online and in physical stores via contactless payments. In recent years, the use of Bitcoin for everyday transactions has increased, as more merchants and retail traders have begun accepting it as a form of payment. You remember how this works as I explained in lesson 2.

Facilitate Global Travel

Bitcoin is a borderless and boundless digital money. It can be easily transferred and converted into local currencies, making it a convenient way to access funds while traveling. There have been incidences of Bitcoin users traveling the world solely relying on Bitcoin payments. You can be sure not to worry about exchange rates when traveling with Bitcoin.

Alternative Store of Value

Bitcoin is used as a good asset for preserving value over the long term, similar to gold. Its value is determined by market supply and demand, rather than being tied to any physical asset or government policy. Increasing adoption in the use of the currency only means increasing demand over time. Any value stored in Bitcoin will most likely increase rather than diminish.

Hedge against Inflation

The value of Bitcoin has the potential to withstand inflation over time. In the early days of Bitcoin, 10,000 bitcoins was used to pay for two pizza boxes. Today, even though the price of a single pizza box has skyrocketed in terms of fiat currency values, 1 Bitcoin today can possibly buy hundreds or thousands of pizza boxes. This is possible because Bitcoin has a limited supply of 21 million coins, making it a deflationary currency. This means that as demand for Bitcoin increases, the value of each coin will also increase, making it a hedge against inflation. Additionally, Bitcoin is not subject to the monetary policies of any central bank or government, so its value is not affected by their decisions to print more money, which can devalue fiat currency.

Collateral for Loans

Bitcoin can be used as collateral for loans because it holds value and is recognized as a legitimate form of currency. Since Bitcoin is a decentralized currency, it is not subject to government control or inflation, which makes it ideal collateral for loans. Bitcoin can be used as collateral for traditional loans or peer-to-peer loans, and the value of the collateral is determined by the current market price of Bitcoin. Using Bitcoin as collateral can potentially pay off the loan when its price skyrockets before the loan repayment is due.

Crowdfunding for charity and social causes

Bitcoin can be used for crowd funding, which is a process of raising funds for a project or cause through small contributions from a large number of people. Crowdfunding through Bitcoin can help charities and social causes by providing a fast and secure way to receive donations from around the world. Bitcoin can be sent directly to the charity's wallet address without any intermediaries, which means that there are no fees or delays involved.

Summary

Bitcoin is a decentralized digital currency that offers many potential uses. It is a low-cost alternative to traditional financial transactions, simplifies cross-border payments, and ensures privacy through pseudonymous user identities. Bitcoin also offers a means to boycott censorship and provides financial inclusion for the unbanked. It has anti-corruption potential due to publicly viewable transaction records, and it can be used to pay for goods and services. Additionally, Bitcoin is a convenient way to access funds while traveling, acts as an alternative store of value and a hedge against inflation, and can be used as collateral for loans. Lastly, Bitcoin can be used for crowdfunding for charity and social causes.

Module B
The Technology behind Bitcoin

In this module, we will be exploring the technical aspects of Bitcoin, including the Bitcoin blockchain, mining processes, nodes, and other key concepts.

The blockchain is the underlying technology that powers Bitcoin and is a decentralized public ledger of all Bitcoin transactions. We will dive into how the blockchain works, its structure, and its security mechanisms.

Bitcoin mining is the process by which new Bitcoin is created and verified. We will discuss the mining process, the hardware and software required, and the role of miners in maintaining the security of the network.

Bitcoin nodes are an essential component of the Bitcoin network, and we will explore how they function and the role they play in the network's security and decentralization.

The Bitcoin halving is an event that occurs approximately every four years, where the block reward for miners is reduced by half. We will discuss the significance of this event and its impact on the Bitcoin ecosystem.

We will also dive into the mechanics of how Bitcoin transactions work, including UTXOs and change addresses. Furthermore, we will examine

Bitcoin forks, which are a divergence in the blockchain resulting in two separate chains with different rules.

Finally, we will explore Bitcoin Improvement Proposals (BIPs), which are a way for the Bitcoin community to suggest and implement changes to the Bitcoin protocol.

By the end of this module, you will have a solid understanding of the technical underpinnings of Bitcoin, allowing you to appreciate its revolutionary potential fully.

Lesson 7
The Bitcoin Blockchain

You've read about the term "Bitcoin Blockchain" several times in the first module of this course. But what is it exactly? In simple terms, the Bitcoin blockchain is a digital ledger that records all Bitcoin transactions.

This ledger is made up of blocks that contain transaction records, and each block is linked to the one before it. The blocks are secured using cryptography, which makes it almost impossible for anyone to tamper with the records.

Imagine there's a big record book in the supermarket close to you. The cashier records all the purchases and sales made in the supermarket in this book. Both suppliers and buyers can always meet the cashier at the counter to record their purchases and deliveries. When this is done, they can ask to have a look to confirm that this is done right and may sometimes be asked to sign against their records in approval. This is similar to how the Bitcoin blockchain works.

But what makes the Bitcoin blockchain so special is that it is decentralized. This means that there is no central authority or intermediary controlling the transactions. Instead, the blockchain is maintained by a network of computers around the world.

Whenever a new transaction is made, it is added to a block by a network of computers called miners. These miners compete to solve a complex mathematical problem, and the first one to solve it gets to add their new block to the blockchain. Each block is like a page in the cashier's record book containing transactions. The miners are the cashiers.

Once a block is added to the chain, it cannot be modified. This is what makes the blockchain resistant to tampering and fraud. It also means that every transaction made on the blockchain is permanent and cannot be reversed.

To ensure the integrity of the blockchain, the network uses a consensus algorithm called proof of work. This algorithm is used to verify and validate new blocks by checking that the transactions contained in the block are valid and that the block is properly linked to the previous block in the chain. This is done by Bitcoin Nodes.

So, to summarize all I've written above, the Bitcoin blockchain is a decentralized ledger that records all Bitcoin transactions. It is made up of blocks, containing transactions, that are secured using cryptography and linked together in a chronological order. Once a block is added to the chain, it cannot be modified, making it resistant to tampering and fraud. The network uses a consensus algorithm called proof of work to validate new blocks and maintain the integrity of the blockchain.

But why is the Bitcoin blockchain so important?

Well, it has the potential to revolutionize the way we do transactions. Because it is decentralized, it eliminates the need for intermediaries like banks or payment processors. This means that transactions can be made faster, cheaper, and more securely than traditional methods. More than anything, it solves the problem of double spending, which we're going to learn about in the next module.

Additionally, because the blockchain is a permanent and transparent ledger, it can help prevent fraud and corruption. It also has the potential to bring financial services to people who are unbanked or underbanked, giving them access to financial services that they may not have had otherwise.

The Bitcoin blockchain is an important innovation in the world of finance and technology. It took the world by storm, offering freedom of transaction to everyone in the world who has access to the internet.

Summary

The Bitcoin blockchain is a digital ledger that records all Bitcoin transactions. It is made up of blocks secured using cryptography and linked together chronologically. The blockchain is decentralized, eliminating the need for intermediaries like banks or payment processors, making transactions faster, cheaper, and more secure than traditional methods. The blockchain's permanent and transparent ledger can help prevent fraud and corruption, and provide financial services to the unbanked or underbanked. The blockchain uses a consensus algorithm called proof of work to validate new blocks and maintain its integrity. The Bitcoin blockchain is an important innovation in finance and technology, offering freedom of transaction to everyone with internet access.

Lesson 8
Bitcoin Mining

Bitcoin mining is a process that sustains the existence of the Bitcoin Network. It involves adding transaction records to the Bitcoin Blockchain. When you initiate a Bitcoin transaction, it is broadcast on the Bitcoin network and goes into the Blockchain's memory pool (mempool). This mempool contains all new Bitcoin transactions waiting to be verified, and all Bitcoin miners have access to it.

Miners collect some of the transactions in the mempool into a block they newly created. This new block with transactions must also contain the digital signature (called hash) of the last block that was successfully added. This is named "previous block hash." The block also contains a nonce, a randomly generated number that must not be repeated twice. With all this information, the block is missing one thing that is needed to make it worthy of being added to the Bitcoin Blockchain - its own hash. This hash must be lower than a specific hash value which the network has fixed by default.

Miners use their available computing power to figure out this new block's hash by constantly making attempts to guess this new hash value by burning lots of energy and computational power. Finding a successful hash value for the new block means the miner did enough work as required by the protocol. This is why this process and resulting hash value is called proof of work. The first miner to find the hash value that is below the target hash wins the arm race.

The successful miner immediately announces their proof of work finding to the network. The other miners immediately check to be sure that the new block only contains valid transactions. If all the conditions are met and this is agreed upon by more than 50% of the overall miners, this new block is added to the Bitcoin Blockchain. The miner whose block was successfully added to the blockchain gets some block rewards. This comprises the new block's incentive in the form of brand new bitcoins and all the fees collected from each of the transactions included in the block.

Mining Bitcoin serves two main purposes: it validates transactions and creates new bitcoin.

Every time a transaction is made, it needs to be verified to ensure that it is legitimate. This is where Bitcoin miners come in. They use powerful computers to verify transactions and add them to the blockchain, which is the digital ledger that records all Bitcoin transactions.

The other purpose of Bitcoin mining is to create new bitcoins. This comes as a form of incentive to miners for their hardwork in solving complex mathematical problems. The rate at which new

Bitcoin is created is reduced by half every 210,000 blocks, which takes roughly 4 years. This is known as the Bitcoin halving.

What does it Cost to mine Bitcoin?

Bitcoin mining is not an easy task. It requires a lot of computational power and resources. In the early days of Bitcoin, it was possible to mine Bitcoin using a personal computer. However, as more people started mining Bitcoin, the difficulty level increased, making it almost impossible to mine Bitcoin with a personal computer.

Nowadays, to be successful in Bitcoin mining, you need to have a specialized device called an Application Specific Integrated Circuit (ASIC). This is specially designed as a Bitcoin mining hardware and can consume a lot of energy both for running and cooling.

Thus, the cost of setting up a Bitcoin mining rig can be quite high. Therefore, instead of doing it alone, many miners join a mining pool. A mining pool is a group of miners who pool their resources together to increase their chances of success. When a member of the pool successfully mines a block, the rewards are shared among all the members of the pool depending on the percentage of their stake in the pool.

As more miners join the network, the difficulty level of Bitcoin mining increases. This is because the more miners there are, the more computing power is required to solve the mathematical problems. The difficulty level is adjusted every 2016 blocks, which takes roughly 2 weeks, to ensure that it takes approximately 10 minutes to mine a block.

Bitcoin mining is an essential process that helps to secure the blockchain and ensure that all Bitcoin transactions are accurate and reliable. When miners successfully verify a block, they receive brand new Bitcoin as a reward. This increases the amount of Bitcoin in circulation every ten minutes.

Summary

To sum it up, Bitcoin mining is the process of verifying Bitcoin transactions and adding them to the blockchain. It involves solving complex mathematical problems using specialized devices called ASICs. Bitcoin mining helps to secure the blockchain and create new bitcoin, which increases the amount of bitcoins in circulation. By understanding Bitcoin mining, you can gain a better understanding of how Bitcoin works and why it is so valuable.

Lesson 9
Bitcoin Nodes

One of the fundamental aspects of Bitcoin is nodes, and this lesson will give you an overview of what they are and their role in the Bitcoin network.

What are Bitcoin Nodes?

Bitcoin Nodes are responsible for keeping the Bitcoin network decentralized, secure, and functional. They are devices that store a copy of the Bitcoin Blockchain software.

These nodes are connected to the entire Bitcoin network and share information with other nodes on the network. They verify that transactions are valid before adding them to the Bitcoin Blockchain, and they keep a detailed record of all transactions to ensure there are no double-spending issues in the protocol.

There are three types of Bitcoin Nodes: Full Nodes, Mining Nodes, and Light Nodes.

Full Nodes

Full Nodes are the most crucial type of node in the Bitcoin network. These nodes download a copy of the entire Bitcoin Blockchain and validate each incoming transaction by checking that it is properly signed and follows the rules of the Bitcoin protocol. Full Nodes then broadcast valid transactions to other nodes on the network and reject invalid transactions and blocks from miners. Full Nodes are essential for maintaining the integrity and security of the Bitcoin network.

Anyone with a good computer with constant internet access and a power source can run a Full Node. However, if you want to mine Bitcoin, you need more than just a Full Node.

Mining Nodes

Mining Nodes are Full Nodes, but with additional capabilities. They participate in the competitive computational activity that creates valid Bitcoin blocks that come with newly generated bitcoins and a collection of verified Bitcoin transactions. Mining Nodes are what we call Bitcoin miners, and they utilize significant computational power and energy to create valid blocks.

Mining Nodes collect verified transactions from the mempool (a list of unspent transaction outputs), and each mining node uses their sophisticated mining hardware to solve the mathematical puzzle. Once they find the needed answer to the puzzle relevant to their new block, they broadcast their new block to the entire network of Full Nodes (including other miners).

Once over half of the Full Nodes have sufficiently verified that the block is valid and contains validated transactions only, they add the new block to the existing copy of the Blockchain in their computers.

While Mining Nodes utilize significant amounts of computational power and energy to create valid blocks, Full Nodes simply verify these blocks without doing any computational work. Thus, we can say that all Mining Nodes are Full Nodes, but not all Full Nodes are Mining Nodes.

Light Nodes

Light Nodes, also known as Simplified Payment Verification (SPV) nodes, do not download a copy of the entire Blockchain. Instead, they only download the block headers, which contain information about the most recent blocks but not the transactions themselves. They rely on Full Nodes to provide them with information about specific transactions, and they verify that the information on the network was validated.

Because they don't need to download the entire Blockchain, SPV nodes can run on devices with limited storage or computational power, such as smartphones. In fact, most self-custody wallets are lightweight nodes. They broadcast specific transactions to the Bitcoin network and check the validity of a connected transaction as relayed to it by Full Nodes.

Light Nodes depend on Full Nodes for their information, and as a result, they don't have full unrestricted access to the Bitcoin Blockchain. If you want full control as a network participant and ensure your access is fully decentralized, you need to run a Full Node. This way, you have full authority over the Blockchain information available to your Bitcoin wallet.

Summary

Bitcoin Nodes are the backbone of the Bitcoin network. Full Nodes ensure that all transactions are valid, Mining Nodes add new blocks to the Blockchain, and Light Nodes allow for lightweight access to the network. All these different types of nodes play a crucial role in maintaining the decentralized, secure, and transparent nature of the Bitcoin network.

By understanding the basics of Bitcoin Nodes, you can make more informed decisions when it comes to managing your Bitcoin wallet, participating in the network, or even contributing to its infrastructure by running your own Full Node.

Lesson 10
Bitcoin Halving

Bitcoin Halving is a process that occurs in the Bitcoin Blockchain every time 210,000 successful blocks are mined. It's a built-in mechanism that reduces the amount of new bitcoin that is created through mining by cutting the mining reward in half. This process is designed to maintain scarcity and prevent inflation by controlling the supply of Bitcoin.

When the mining reward is reduced, the supply of Bitcoin is constrained, which can have a positive impact on its price if demand remains the same. For instance, if there are fewer Bitcoins available, but the same number of people want to buy them, the price of Bitcoin will go up.

How often does Bitcoin Halving occur?
Bitcoin Halving takes place every 210,000 successful blocks mined. Since it takes about 10 minutes to mine each block, this means that Bitcoin Halving occurs roughly every 4 years. This can be less or more eventually.

When did the first Bitcoin Halving occur?
The first Bitcoin Halving occurred in November 2012. This was followed by another halving in July

2016, and the most recent halving took place in May 2020. Each time a new halving occurs, the mining reward is cut in half. For example, after the first halving, the reward for mining was reduced from 50 BTC per block to 25 BTC per block. The second halving cut that in half giving 12.5 BTC per block. The most recent halving reduced the reward further to 6.25 BTC per block, which is the current subsidy paid to miners.

Why is Bitcoin Halving important?

The goal of Bitcoin Halving is to maintain the scarcity and finiteness of Bitcoin by controlling its issuance. The total supply of Bitcoin is fixed at 21 million, and as at March 2023, a bit more than 19 million have been mined. This means that there are a little less than 2 million bitcoins left to be created.

By reducing the number of new coins issued with each new block, Bitcoin can maintain its scarcity and prevent inflation.

Why does this matter? Well, one of the biggest concerns with traditional currencies is inflation. As governments print more money, the value of that money decreases, and prices go up. This can be a problem for people who are trying to save money or protect their wealth. Bitcoin's scarcity is designed to counteract this. By limiting the number of bitcoins that can be created, it helps to maintain their value over time.

The halving process is a key part of this. By decreasing the mining reward, it helps to slow down the rate at which new bitcoins are created. This makes Bitcoin a more attractive investment option for people who are concerned about inflation.

What is the impact of Bitcoin Halving on mining profitability?

The halving process also has an impact on mining profitability. When the reward is cut in half, some miners may find it no longer profitable to continue mining. This can lead to a reduction in the overall hashrate and security of the network.

Despite this, the halving process is generally seen as a positive for the Bitcoin ecosystem. It helps to maintain the scarcity and finiteness of Bitcoin, which in turn helps to support its value. It's worth noting that the halving process will continue until the year 2140, at which point all 21 million bitcoins will have been mined.

Summary

Bitcoin halving is an important process that helps to maintain the scarcity and value of Bitcoin. It happens every 210,000 blocks mined and involves cutting the mining reward in half. This helps to slow down the rate at which new bitcoins are created and makes Bitcoin a more attractive investment option for people concerned about inflation. While it can have an impact on mining profitability, the halving process is generally seen as a positive for the Bitcoin ecosystem.

Lesson 11
How Bitcoin Transactions Work

Bitcoin uses encryption to verify transactions and control the creation of new units. Unlike traditional currencies, it is not managed by any central authority such as a government or a bank. Instead, Bitcoin transactions work by sending digitally signed messages across the Bitcoin network for verification.

Each Bitcoin transaction is like an email that is digitally signed using cryptography and broadcasted to the entire Bitcoin network for verification. The transaction information is available on the Bitcoin Blockchain, which is a public ledger for all to see. Each transaction record is arranged in chronological order.

Bitcoins are not physical and users do not need to have an "account." Instead, there is only the record of transactions on the blockchain that have been sent from one address to another. When users want to send Bitcoin, they need to have access to the public and private keys associated with the amount of Bitcoin they want to send.

What it means to own Bitcoin

When someone "owns" Bitcoin, it means they have access to a 'key pair' that consists of a public key (an address) to which some amount of Bitcoin was previously sent, and the corresponding unique private key (a password) that authorizes the Bitcoin previously sent to be sent elsewhere. Public keys, also called Bitcoin addresses, are randomly generated sequences of letters and numbers that function similarly to an email address or a social media username. They are public, so it is safe to share them with others. In fact, users must share their Bitcoin address with others when they want them to send Bitcoin. Private keys are also randomly generated sequences of letters and numbers, but they are to be kept secret like passwords to email or other accounts.

Users can think of their Bitcoin address as a transparent safe. Others can see what's inside, but only those with the private key can unlock the safe to access the funds within.

Sending the Bitcoin you own

To send Bitcoin, users need to sign a message with their private key that contains transaction-specific details. This message must be broadcast to the network and contains information about the Bitcoin previously sent to the user's address, the amount the user wants to send, and the address of the recipient.

For example, let's say Mark wants to send 1 Bitcoin to Jessica. He uses his private key to sign a message with the transaction-specific details, including the inputs (information about the Bitcoin previously sent to Mark's address), Jessica's address, and the outputs (the amount Mark wants to send to Jessica and the amount returned to Mark as 'change,' if any). This message is then broadcast to the Bitcoin network.

Miners Verification of the ownership and Confirmations

A special group of participants in the network known as 'miners' verify that Mark's keys are able to access the inputs (i.e. the address(es) to where he previously received the Bitcoin he claims to control.) Miners also gather together a list of other transactions that were broadcast to the network around the same time as Mark's and form them into a block. Any miner who has completed the 'Proof of Work' is permitted to propose their new block which will be added or 'attached' to the blockchain if valid. The new block must reference the latest old one. That new block is then broadcast to the network. If other network participants (full nodes) agree it's a valid block (ie. the transactions it contains follows all the rules of the protocol and it properly references the previous block), they will pass it along.

When over half have approved this new block, the transactions are appended on the Blockchain permanently. As more blocks are added to the chain, the number of confirmations of Mark's transaction increases. Jessica can then check to confirm that she got the Bitcoin sent by Mark. Mark can also see his change, if any.

Why some transaction confirmations may take longer

As a beginner in the world of crypto, you may be wondering why some Bitcoin transaction confirmations take longer than others. The answer lies in the limited space available in each block, which can only contain a certain number of transactions. This limit, also known as the block size, is currently set at 1MB.

Because of the limited space, there is a fee market where miners choose to first include those transactions with high enough fees in the next block. This means that higher fees act as an incentive for miners to prioritize your transactions. It's important to note that the block size is not set in stone and is an arbitrary limit. However, the Bitcoin community has chosen to keep it small to make it easier for people to operate Bitcoin nodes.

How costly can Transaction Fees be?

If you're wondering how much Bitcoin transaction fees are, they can vary widely, from just a few cents up to $100 or more. The reason for this variation is that Bitcoin fees depend on both supply and demand, which means that the more congested the network is, the higher the fee will be. Additionally, the "size" of your transaction can also affect the fee. If your transaction has many inputs, it will take up more block space and demand a higher fee. For example, if you want to send 10 BTC received at once to your address, your transaction has only one input. On the other hand, if you were sending 1 BTC sum of which was received at three different times, your transaction will have 3 inputs. This way, sending 10 BTC might cost less than when you want to send your 1 BTC. I'll discuss this in detail in the next lesson.

Thankfully, many wallets allow you to manually set your transaction fees, which can help you avoid overpaying. If you're not in a rush, you can set the fee lower and wait for the network to be less congested. On the other hand, if you need your transaction to be processed immediately, you can increase the fee.

Overall, while the process of sending and receiving Bitcoin may seem complex, it's typically done through a Bitcoin Wallet software. This software allows you to create and manage your public and private keys, send and receive Bitcoin, and view your transaction history. You don't have to understand how everything works at the backend as explained in this lesson.

Summary

Bitcoin is a decentralized digital currency that uses encryption to verify transactions and control the creation of new units. Each transaction is like an email that is digitally signed using cryptography and broadcasted to the entire Bitcoin network for verification. When someone owns Bitcoin, it means they have access to a key pair consisting of a public key and a unique private key. To send Bitcoin, users need to sign a message with their private key that contains transaction-specific details, which is then broadcast to the network. Miners verify the transaction and form it into a block. The number of confirmations of a transaction increases as more blocks are added to the chain. Transaction confirmations can take longer due to the limited space available in each block, and fees can vary depending on demand and transaction size.

Lesson 12
UTXOs and Change Addresses

In this lesson, we're going to talk about Bitcoin UTXOs and change addresses. If you're new to Bitcoin cryptocurrency, this may sound confusing, but don't worry! I'll explain it in a way that's easy to understand.

To start, let's talk about Bitcoin transactions. When you send or receive Bitcoin, it involves transferring value from one or more Bitcoin addresses to another. Each transaction consists of three parts: inputs, amount to send, and outputs.

Inputs and outputs are both part of a Bitcoin UTXO, which stands for Unspent Transaction Output. Inputs are UTXOs that you are spending in a transaction, because you own them. Outputs are new UTXOs that you are crediting to the receiver's address, and to your own newly generated address if you have change.

So, what exactly are UTXOs?

Simply put, they are denominations of Bitcoin or satoshi that you have received in your address per transaction. For example, if you send 0.1 BTC from your MEXC address to your non-custodial wallet address, you have a UTXO worth 0.1 BTC. If someone else sends you 0.3 BTC to the same address, you now have a second UTXO in the denomination of 0.3 BTC. Each Bitcoin transaction you receive in your Bitcoin address generates a new UTXO, and all these UTXOs together form your Bitcoin wallet address balance.

Each UTXO exists separately from the others in your address. This is because Bitcoin does not operate on an account system like most other cryptocurrencies. Instead, every piece of information about Bitcoin exists as a set of UTXOs on the Bitcoin blockchain.

Thus, the UTXO model stores each transaction received in any address in separate denominations.

An allegory

A good way to understand this is by thinking about cash in a piggy bank. Let's say you have a piggy bank where you input $1, $5, $20, or even $100 bills per day, depending on what you can afford to keep away each day. Your piggy bank can have different denominations depending on which dollar bill you put into it on any particular day. After 365 days, let's say you have a total of $15,000 in your piggy bank. This amount does not exist in a single denomination like a $15,000 bill. Instead, if you open the piggy bank, you can see all the different dollar bills in the same way you put them. None of your $20 bills magically became $100 bills.

The same applies to Bitcoin transactions. The only difference is that Bitcoin has no specific denominations. A UTXO denomination is simply any amount of BTC received per transaction.

Inputs and Outputs

When any transaction is made on the Bitcoin blockchain, the funds being transferred are taken from one or more UTXOs as inputs and are sent to one or more new addresses as outputs. Your inputs are Bitcoin UTXOs you have the right to spend because you can access them from your private keys. The outputs are Bitcoin UTXOs you send to others and yourself (if you have change).

Here's an example to make it clearer. Imagine that you have a Piggy bank with one $100 and two $20 dollar bills. You are going to a shop next to your house to buy something that costs $130. However, you do not have the exact amount in denominations; you have one $100 bill and two $20 bills. You give the cashier all three bills, totaling $140, and the cashier returns $10 in change. This is how UTXOs work.

The $100 and two $20 bills are your transaction inputs (3 in all), while the $130 spent and your change of $10 are the outputs.

Just like when you pay for something with cash, you need to have the right amount of Bitcoin to complete a transaction. In Bitcoin transaction, we use "unspent transaction outputs" (UTXOs) to represent the value that can be spent in a transaction. These UTXOs are created as a result of previous transactions, and they are destroyed when they are spent in a new transaction, while creating new ones in the process.

For example, if you received three separate transactions of 0.1, 0.4, and 0.6 bitcoins and wanted to send 0.8 bitcoins to someone else, your existing UTXOs of 0.1, 0.4, and 0.6 would be combined as the inputs for the transaction.

The outputs of the transaction are where the bitcoin is going. In this case, you would have two outputs: 0.8 bitcoins to the person you're paying and 0.3 bitcoins back to you as change. This is similar to how the cashier gave you $10 back as change.

UTXOs storage

UTXOs are stored in a database called the UTXO set, which contains all unspent transaction outputs on the network. When you make a transaction, you need to have access to one or more UTXOs to use as inputs. The sum of the inputs must be greater than or equal to the sum of the outputs (the amount you're sending plus any change you're getting back). The difference between the inputs and the outputs is called the transaction fee.

Change Addresses

Now, let's talk about change addresses. When you make a transaction, you might have more cryptocurrency in your possession than you need to complete the transaction. For example, if you have 1 bitcoin and you want to send 0.5 bitcoin to someone, you'll have 0.5 bitcoin left over. This left-over amount is called "change," and it needs to be sent back to your own address. This is where a change address comes in.

A change address is a new address that is used as the destination for any change you receive from a transaction. This address is usually different from the address you're sending from, but it's still part of your wallet balance because you own the private key. The use of change addresses is an important feature of Bitcoin transactions. It ensures that your remaining funds are not lost and can be used in future transactions. Additionally, it enhances the privacy and security of transactions by making it more difficult for outside parties to track the flow of funds.

But don't worry - you don't need to understand UTXOs and change addresses in detail. When you make a transaction, the software that creates the transaction (like your wallet) will automatically select the UTXOs to be spent and determine the change address. The change address is usually determined by the software wallet, but it can also be manually set by the user.

Summary

Understanding UTXOs and change addresses is important for cryptocurrency transactions, but it's not necessary to know all the technical details. Just remember that UTXOs represent the value that can be spent in a transaction, and change addresses are used to send any leftover cryptocurrency back to your own address. With this knowledge, you're ready to start making your own Bitcoin transactions!

Lesson 13
Bitcoin Forks

A Bitcoin fork is basically a change to the software that powers Bitcoin. This change can split the Bitcoin network into two separate chains, each with its own set of rules. When this happens, a new cryptocurrency can be created alongside the original Bitcoin. This is because the new software protocol introduces new features or changes how the network already works, which may appeal to some network participants.

Types of Bitcoin Forks

There are two types of Bitcoin forks: soft and hard forks.

Soft forks are changes to the software that are backwards-compatible. This means that nodes running the new software can still communicate and validate transactions with nodes running the old software. Soft forks typically introduce new features or improve existing ones, and they tend to be less contentious since they don't involve a complete overhaul of the protocol.

On the other hand, hard forks are not backwards-compatible and require all nodes to upgrade to the new software. This type of fork can result in the creation of two separate cryptocurrencies if a portion of the network chooses not to upgrade, as happened in the case of Bitcoin Cash. Hard forks can also result in a chain split if the majority of the network is not in agreement on the changes being implemented, leading to two separate and incompatible chains.

History of Bitcoin Forks

Throughout the history of Bitcoin, there have been several notable forks. For example, Bitcoin Cash was created in 2017 when a group of developers and miners proposed a hard fork to increase the block size limit from 1 MB to 8 MB in order to improve scalability. The hard fork resulted in the creation of Bitcoin Cash, which has since become one of the largest cryptocurrencies by market capitalization.

Another notable example is Bitcoin Gold, which was created in 2017 when another hard fork was proposed to change the proof-of-work algorithm from SHA-256 to Equihash, making it easier for individuals to mine the cryptocurrency using graphics processing units (GPUs).

In August 2017, a soft fork was implemented to improve scalability by separating signature data from transactions and storing it in a separate structure called a witness. This allowed for a higher degree of transaction throughput and reduced transaction fees. This is what Bitcoin uses today and is called SegWit.

Bitcoin Satoshi's Vision (BSV) is a type of cryptocurrency that was created in November 2018 as a result of a Bitcoin hard fork. This hard fork was proposed to increase the block size limit to 128 MB and restore the original Bitcoin protocol as outlined by Satoshi Nakamoto in the original Bitcoin white paper.

Forks in general have played a significant role in the development of the Bitcoin network and have also contributed to the creation of other successful cryptocurrencies. However, they have also led to controversy and debate within the community, and they continue to pose challenges to the decentralization and stability of the network.

Summary

A Bitcoin fork happens when there is a change made to the underlying software protocol of the network, which can cause the network to split into two chains, each with its own set of rules. These changes can occur when network participants introduce new features that change how the network operates. There are two types of Bitcoin forks: soft (which are backwards-compatible changes) and hard (which are not backwards-compatible).

Lesson 14
BIPs

Bitcoin Improvement Proposals (BIPs) are formal proposals for making changes and improvements to the Bitcoin network and its software. They are created by developers and members of the community, and once approved by the rest of the Bitcoin community, they become part of the Bitcoin codebase.

BIPs allow for open discussion and collaboration among stakeholders, making the decision-making process decentralized and democratic. Anyone can propose a BIP, and the community can review, discuss, and vote on its implementation. BIPs that are not popular with the majority of the community are often discarded or reworked to meet demands.

Notable BIPs

Some notable BIPs in the history of Bitcoin include SegWit (Segregated Witness), Version Bits, and Bech32.

Segregated Witness (SegWit) was proposed as a BIP in 2015 by Bitcoin Core developer Pieter Wuille. SegWit was a major protocol upgrade that sought to increase the block size limit (to 4megabyte[1] in terms of transaction weight units) and

---
[1] Note that Bitcoin's block size remains 1 megabyte by default.

improve transaction efficiency. The idea was to separate the digital signature data from the transaction data, allowing for more transactions to fit in each block. By doing so, it aimed to reduce transaction fees and improve the overall performance of the Bitcoin network. SegWit was activated on the Bitcoin network in August 2017, and its adoption has continued to grow since then.

Version Bits was proposed as a BIP in 2016 by Bitcoin Core developer Johnson Lau. It was a proposal for a soft fork mechanism that allows for multiple backward-compatible protocol upgrades to be activated at the same time. Version Bits enables miners to signal their support for different proposals using a bit flag in the block header. This allows for a more flexible and efficient process for upgrading the Bitcoin protocol compared to previous methods that required a hard fork.

Bech32 was proposed as a BIP in 2017 by Bitcoin Core developers Pieter Wuille and Greg Maxwell. It was a new format for Bitcoin addresses that improves error detection and makes it easier for users to send and receive Bitcoin. Bech32 addresses are designed to be more user-friendly than previous address formats, such as Base58. They are also more efficient in terms of space usage, allowing for more transactions to fit in each block. Bech32 addresses start with "bc1" and are becoming increasingly popular among Bitcoin users and service providers.

Types of BIPs

There are several types of BIPs, including Standard, Informational, Process, Meta, Interface, Economic, Consensus, and Deployment BIPs.

Standard BIPs describe changes to the core Bitcoin protocol, including new features, changes to existing functionality, and security improvements.

Informational BIPs provide information or documentation on a particular aspect of the Bitcoin network, without proposing any specific changes to the actual protocol.

Process BIPs describe changes to the process for creating and implementing BIPs, such as changes to the voting process or the criteria for determining whether a BIP is accepted or rejected.

Meta BIPs provide a framework for creating and organizing other BIPs, such as proposals for new categories of BIPs or guidelines for BIP authors.

Interface BIPs describe changes to the user interface and user experience of the Bitcoin network, such as new features for users, or improvements to the way transactions are displayed or processed.

Economic BIPs propose changes to the economics of the Bitcoin network, such as changes to the mining reward structure or the issuance of new bitcoins.

Consensus BIPs propose changes to the consensus mechanism of the Bitcoin network, such as changes to the proof-of-work algorithm or the way blocks are validated.

Deployment BIPs describe how and when a particular BIP will be deployed, such as specific timelines for the activation of a new feature, or a plan for gradually rolling out a change to the network.

BIPs go through several stages before they are implemented, including proposal, discussion, drafting, review, and acceptance. Once a BIP is accepted, it becomes part of the official Bitcoin codebase, and its changes are integrated into the network.

Summary

BIPs are suggestions for making changes and improvements to the Bitcoin network and its software. They are created by developers and members of the community and are subject to open discussion and collaboration among stakeholders. There are several types of BIPs, including Standard, Informational, Process, Meta, Interface, Economic, Consensus, and Deployment, each of which describes different aspects of the Bitcoin protocol. BIPs go through multiple stages before implementation and play a crucial role in the decentralized and democratic evolution of the Bitcoin network.

Module C
Bitcoin Network Security

In this module, we will be exploring the various security measures that are in place to ensure the safety and integrity of the Bitcoin network.

Throughout this module, you will gain an understanding of the technical aspects of Bitcoin's security, including the consensus algorithm, cryptography, and decentralization. You will also learn about the potential risks and threats to the Bitcoin network, such as the 51% attack and double spending, and how they can be mitigated.

Furthermore, this module will explore the concept of privacy in Bitcoin and how it can be achieved through various techniques such as mixing and tumbling. You will also learn about the importance of decentralization in Bitcoin and how it contributes to its security and resilience.

Lastly, you will discover why Bitcoin consumes so much energy and the various arguments for and against its energy consumption. By the end of this module, you will have gained a comprehensive understanding of Bitcoin's security measures and how they contribute to the overall stability of the network.

Lesson 15
Bitcoin's Security

As a digital currency that is not controlled by any central authority or intermediary, Bitcoin uses a technology called blockchain to record all transactions in a tamper-proof and secure way. How does this work?

To ensure the security of the Bitcoin network, two important mechanisms are used: cryptography and proof-of-work. These two lead to the concepts of game theory and decentralization.

Cryptography

Cryptography is a technique used to keep information secret and secure by the use of encryption techniques. In the case of Bitcoin, cryptographic algorithms are used to secure all incoming Bitcoin transactions and prevent anyone from altering the data already stored in the blockchain. When someone makes a transaction on the Bitcoin network, their transaction is verified and signed using a complex cryptographic algorithm. This signature acts as a digital seal that ensures the integrity and authenticity of the transaction. It proves that the transaction was initiated by the real owner of the Bitcoin being spent.

Proof of Work

Proof-of-work is a consensus mechanism used in the Bitcoin network to validate transactions and maintain the security of the blockchain. In order for a transaction to be added to the blockchain, it needs to be validated by a network of nodes, known as miners. Miners are responsible for verifying transactions and adding them to the blockchain. They do this by solving complex mathematical problems that require a significant amount of computational power (Check module 2 for details on this).

Proof-of-work ensures the security of the Bitcoin network by making it extremely difficult to alter the blockchain. This is because any changes to the blockchain would require a massive amount of computational power, which is infeasible for a single person or entity to achieve. In addition, because miners are incentivized to behave honestly, they are motivated to follow the rules of the network and prevent any fraudulent activities. This is known as Game Theory in Bitcoin.

Game Theory

Bitcoin's security is not just based on cryptography and proof-of-work, but also on game theory. Game theory is the study of strategic decision-making involving more than one subject, and it plays a crucial role in Bitcoin's consensus mechanism.

In Bitcoin, miners compete to solve a cryptographic puzzle, and the first one to solve it gets to add a new block to the blockchain and receive a reward in bitcoins. However, since there are many miners, they are also competing with each other. This competition creates a game-theoretic situation called the "mining game."

In the mining game, miners are incentivized to behave honestly and follow the rules of the Bitcoin protocol, because if they don't, they risk losing their investment in mining equipment and electricity. However, there is also a temptation to cheat and try to manipulate the blockchain to their advantage.

To prevent cheating, the Bitcoin protocol is designed to make it very difficult and expensive to cheat. For example, if a miner tries to add an invalid block to the blockchain, other miners will reject it, and the miner will not receive a reward. This creates a disincentive to cheat, because the potential rewards are not worth the risk of losing the investment.

The mining game also creates a balance of power between miners. If one miner gains too much computational power, other miners will be incentivized to join forces to prevent that miner from controlling the network. This creates a game-theoretic situation known as a "Nash equilibrium," where all miners have an incentive to follow the rules and not try to cheat the system.

Decentralization

Decentralization is a key feature of the Bitcoin network and is one of the primary reasons why it is considered to be secure. Decentralization refers to the fact that the Bitcoin network is not controlled by any single entity, such as a government or financial institution. Instead, it is maintained by a network of users who work together to validate transactions and keep the network secure.

Because there is no single point of control in the Bitcoin network, it is difficult for any one person or group to manipulate the system for their own gain. In a centralized system, such as a traditional bank, there is a single point of control that can be targeted by attackers. This makes it easier for hackers or other malicious actors to compromise the security of the system. However, in a decentralized system like Bitcoin, there is no single point of control, making it much more difficult for attackers to compromise the security of the network.

Overall, decentralization is a key factor in the security of the Bitcoin network. By eliminating the need for intermediaries and creating a network of users who work together to validate transactions, Bitcoin is able to create a secure and decentralized system that is resistant to attacks and manipulation.

Potential Risks

One major security concern for Bitcoin is the potential for "51% attacks," where a group of miners or malicious actors can gain control of more than 50% of the network's computational power. With this level of control, they could potentially manipulate the blockchain and double-spend bitcoins. However, this is considered to be a highly unlikely scenario due to the high cost of acquiring the necessary computational power.

Additionally, there is also a risk of "phishing" attacks against Bitcoin users themselves. An attacker may send an email or message that appears to be from a legitimate source, but is actually a trap designed to steal personal information or login credentials. To avoid this, users should be careful where you enter login details for your Bitcoin accounts on exchanges, and never share your self-custody wallet seed phrases with anyone, friend or stranger.

It's also important to keep in mind that while the Bitcoin network itself may be secure, the exchanges and platforms that allow users to buy and sell bitcoins are vulnerable to hacking and theft. For this reason, it is crucial to use a reputable exchange and to take steps to secure one's own bitcoins, such as keeping them in a hardware wallet or using multisignature technology.

Summary

Bitcoin is a digital currency that uses blockchain to record transactions. The security of the Bitcoin network relies on cryptography and proof-of-work. Cryptography ensures the integrity and authenticity of transactions by making them tamper-proof, while proof-of-work adds new transactions to the blockchain by solving complex mathematical problems. Game theory and decentralization also play important roles in securing the Bitcoin network. Decentralization makes it difficult for any single entity to manipulate the system, while game theory incentivizes miners to follow the rules of the Bitcoin protocol. The potential risks include phishing attacks and the potential for 51% attacks.

Lesson 16
51% Attack

51% is when a single group or entity controls over 50% of the network's computational power or hashrate, giving them the ability to mess with the network and potentially do bad things like double spending or denial of service attacks.

It's important for the Bitcoin network to be decentralized, meaning no one group or entity has too much control over it. But if a bad group or entity can get a lot of computational power, they can potentially do a 51% attack and take over the network. However, this is very difficult and almost impossible to do.

One reason why a 51% attack is unlikely to happen is because the network is so big. To pull off such an attack, the attacker would need to invest a lot of resources and capital, making it an unprofitable venture. The Bitcoin protocol also rewards honest miners, so there is little incentive to try and pull off a 51% attack.

The Bitcoin protocol uses Proof of Work (PoW) to reach a consensus among network participants. Miners solve complex math equations to add new blocks to the blockchain and get rewarded with new Bitcoins. To do a 51% attack, an attacker would need to control more than 50% of the network's mining hashrate, allowing them to mess with the blockchain and potentially double spend.

Double spending is when an attacker sends their own coins to another account and then reverses the transaction so the coins come back to them. This would allow the attacker to spend the same coins multiple times, but they cannot create Bitcoin from thin air.

A 51% attack could also take the form of a denial of service (DoS) attack, where the attacker prevents other network participants from confirming transactions or producing new blocks. This would give the attacker a monopoly on the blockchain and allow them to control the network. However, they wouldn't be able to steal other people's Bitcoins or stop transactions from being broadcast to the network.

Punishments for a 51% attacker can be severe. They would no longer be able to access rewards for mining new blocks.

The chance of a 51% attack happening on a blockchain network depends on how many people are participating in the network. The more people involved, the less vulnerable the network is to attacks. This is because it would require a lot of resources and capital to control over 50% of the network's computational power. This is why a 51% attack on the Bitcoin protocol is nearly impossible and unprofitable for anyone to try.

Summary

A 51% attack on the Bitcoin network is when a single group or entity controls over 50% of the network's computational power, giving them the ability to potentially do bad things like double spending or denial of service attacks. However, this is highly unlikely because the network is so large, and it would require a lot of resources and capital. The Bitcoin protocol rewards honest miners, providing little incentive to attempt such an attack. Additionally, punishments for a 51% attacker can be severe. The likelihood of an attack depends on the number of people participating in the network, and the more people involved, the less vulnerable the network is to attacks.

Lesson 17
Double Spending in Bitcoin

Before Bitcoin came along, there were a lot of attempts to create digital currencies that didn't rely on central authorities. But they all ran into the same problem: double spending. Double spending is when someone tries to spend the same money twice. It's like counterfeiting with physical money, but for digital currencies.

In the traditional financial system, banks and other intermediaries prevent double spending by keeping track of users' transactions and checking to make sure they have enough money in their accounts before allowing them to spend it. But this system can be manipulated by these intermediaries, which defeats the purpose of decentralization.

That's why Bitcoin was created. It's a decentralized payment system that doesn't rely on intermediaries. Instead, it uses the blockchain technology to prevent double spending.

When we talk about double spending in Bitcoin, we mean the ability for someone to spend the same Bitcoin multiple times. This is possible because Bitcoin transactions are recorded on a public ledger called the blockchain, but the transactions themselves are not physically transferred between individuals. So, someone could pretend to send their

Bitcoin to someone else, and then still send the same Bitcoin to themselves or other people.

But this problem is solved by a consensus mechanism called proof-of-work. Miners compete to solve a complex math problem in order to group transactions into blocks and add them to the blockchain. The first miner to solve the problem gets to add the block of transactions and is rewarded with newly minted Bitcoins.

Proof-of-work makes double spending much harder, but not impossible. In order to double spend, an attacker would need to control more than 50% of the mining power on the network, which is known as a 51% attack (explained in previous lesson). If an attacker can control 51% of the mining power, they can create a longer blockchain than the one the rest of the network is working on, which allows them to spend the same Bitcoins multiple times.

However, the distributed nature of the Bitcoin network makes a successful 51% attack very unlikely. There are thousands of miners all over the world working to validate and record transactions on the blockchain. It would take a lot of resources to control 51% of the mining power, making it impractical and uneconomical for an attacker to try.

One way to stay safe from being deceived with a double spending is to wait for several confirmations before finalizing a transaction. When a miner adds a block of transactions to the blockchain, it's considered to be one confirmation. Each subsequent block that is added to the blockchain is considered an additional confirmation. The more confirmations a transaction has, the harder it is to alter or reverse the transaction. Waiting for 6 confirmations, which takes about an hour, is considered safe.

Bitcoin miners also use a mechanism called "first-seen rule" to prioritize the first transaction that is seen by the network. When a miner receives a new transaction, it's broadcast to other nodes in the network. If another miner sees two conflicting transactions, it will only accept the one it saw first. This prevents double spending because the miner will only accept the first transaction, which is considered legitimate, and won't accept any subsequent conflicting transactions involving the same bitcoins.

Summary

Before Bitcoin, digital currencies had a double spending problem, where someone could spend the same money twice. Bitcoin uses blockchain to prevent double spending by using a consensus mechanism called proof-of-work, where miners compete to group transactions into blocks and add them to the blockchain. It would be hard to carry out a 51% attack to control the network and double spend. Waiting for several confirmations before finalizing a transaction and the "first-seen rule" also prevent double spending.

Lesson 18
Privacy in Bitcoin

Bitcoin allows users to send and receive payments anonymously without the need for a middleman, such as a bank. However, despite being anonymous, Bitcoin transactions are still traceable. In this lesson, we will explore how Bitcoin protects users' privacy and what measures you can take to keep your transactions anonymous.

Bitcoin's Use of Pseudonymous Identities

Bitcoin is unique because it does not require users to share personal information when they create a Bitcoin address. Instead, they are assigned a pseudonym, which is like a nickname that is made up of a long string of letters and numbers. The pseudonym is not linked to any real-life identity, so users can remain anonymous if they choose to.

In fact, users can create as many pseudonyms as they want. This makes it difficult for anyone to link multiple transactions to a single person. For example, imagine that you want to buy a pizza using Bitcoin. You could create a new pseudonym for that transaction, and then never use it again. No one would be able to trace that transaction back to you unless you chose to reveal your identity.

The use of pseudonyms is important for protecting user privacy. In traditional banking systems, users are required to share personal information, such as their name, address, and social security number, when they open an account. This information can be used to track and monitor a user's financial activity, which can be a privacy concern.

With Bitcoin, users have more control over their personal information. They can choose to reveal as much or as little as they want, and they can use multiple pseudonyms to keep their transactions private. While this level of privacy may be attractive to some users, it can also be a challenge for law enforcement agencies who are trying to track down criminals who use Bitcoin for illegal activities.

User Responsibility for Privacy

The transparency of the Bitcoin network means that anyone can see every transaction that takes place on it. Even if you don't know the identity of the person behind a Bitcoin address, you can still see what they're doing with their money.

This transparency means that users need to be cautious about their own privacy. The level of anonymity a user can enjoy depends on how they send and receive transactions from their wallet. For example, if a user buys Bitcoin from a KYC/AML compliant exchange and then sends it to their self-custody wallet, the exchange can associate the address with the user's identity. This would unmask the ownership identity of the address and compromise the user's privacy. Therefore, users must be mindful of their actions when using Bitcoin.

It is essential to understand that Bitcoin leaves the responsibility of privacy in the hands of the users themselves. Users need to take extra steps to ensure their anonymity, such as creating a new wallet address for each transaction, and not revealing their identity when making transactions. Failure to take these steps could make a user's financial information public, which could lead to high risk of attack, even identity theft. Therefore, users must prioritize privacy and take proactive measures to protect their financial information.

More Ways to Protect Your Privacy on Bitcoin

Another way to protect your privacy on Bitcoin is to use a change address. When you make a transaction, any remaining bitcoins in your wallet are sent to a new address, known as a change address. This helps to obscure the flow of Bitcoins and makes it more difficult to track the origin or destination of a particular transaction.

Another way to further anonymize Bitcoin transactions is through the use of mixers. These are services that take in a large number of Bitcoins from multiple users and then send them back out in a different order, making it difficult to trace the original source of the Bitcoins. However, it is important to note that these services can also be used for illegal activities, such as money laundering, and may not be fully anonymous as users are trusting a third-party to not store their identity.

It is also important to note that the use of mixers can be risky. If a mixer is compromised, it can expose the identities of its users. Furthermore, using mixers may raise suspicion and draw attention to your transactions.

Using a change address and a mixer can help to protect your privacy on Bitcoin. However, it is important to be cautious and to fully understand the risks involved in using these methods. As with any financial transaction, it is important to do your research and take steps to protect yourself.

Summary

Bitcoin allows users to send and receive payments anonymously without the need for a middleman, such as a bank. However, Bitcoin transactions are still traceable. Bitcoin uses pseudonymous identities, making it difficult for anyone to link multiple transactions to a single person. Users can create as many pseudonyms as they want to keep their transactions private. It is also the user's responsibility to take extra steps to ensure their anonymity, such as creating a new wallet address for each transaction, and not revealing their identity when making transactions. Using a change address and a mixer can help to protect your privacy on Bitcoin, but there are risks involved. It is important to do research and take steps to protect yourself.

Lesson 19
Decentralization in Bitcoin

Decentralization is a key feature of Bitcoin that sets it apart from traditional financial systems. To understand decentralization, let's first look at how traditional financial systems work. In a traditional system, there is usually a central authority or institution, such as a bank or government that controls and regulates the system. This central authority makes decisions on behalf of the system and has the power to manipulate it.

In contrast, Bitcoin is decentralized, meaning that there is no central authority or control over the network. Instead, it is run by a network of users, or nodes, who all have an equal say in decision making, development and direction of the network. This means that no one person or entity has control over the network.

One of the key benefits of decentralization is increased security. In Bitcoin, the network is distributed across multiple nodes, making it more difficult for a hacker to take control of the entire network. For this to happen, the hacker would need to hack into the computer of at least 51% of all participants, which is almost impossible.

Decentralization also allows for greater transparency and accountability. All Bitcoin transactions are recorded on its public ledger, which means that everyone can view them. This makes it difficult for anyone to spend mutually held funds in bitcoin without everyone noticing.

Another benefit of decentralization is that it eliminates the need for central intermediaries, such as banks or other financial institutions. This means that users can transact directly with each other without the need for a trusted third party. It also eliminates the counterparty risks inherent in traditional financial systems.

Decentralization also means that there is no central point of failure. A single node going down does not bring the entire Bitcoin network down. This is why the Bitcoin protocol has suffered no downtime in its entire existence.

Perhaps most importantly, decentralization enables greater financial inclusion and innovation. Anyone can join the Bitcoin network and use it without seeking approval from any central authority. This means that even people who do not have access to traditional financial services can participate in the global economy.

Overall, decentralization is a key aspect of Bitcoin that provides increased security, transparency, and financial inclusion.

Summary

Bitcoin is decentralized, meaning it has no central authority or control. It's run by a network of users, increasing security, transparency, and accountability. Decentralization eliminates the need for central intermediaries, allowing users to transact directly with each other. It also enables greater financial inclusion and innovation.

Lesson 20
Why Bitcoin Needs so much Energy

One aspect of Bitcoin that has come under scrutiny is its energy consumption. In this lesson, we'll take a closer look at why Bitcoin uses so much energy.

When Bitcoin first started out, its energy consumption wasn't considered a big deal. However, as the popularity of the network grew, so did its energy consumption. Critics argue that Bitcoin consumes more energy than entire countries, and that this is harmful to the environment, especially when non-renewable sources are used.

The energy consumption is mainly due to the use of high-powered computers that compete to solve complex mathematical problems, which is referred to as proof-of-work (PoW) consensus algorithm. Bitcoin miners who solve the problem gets to add a block to the Bitcoin blockchain.

The energy consumption of Bitcoin has been the subject of much debate. Those who support Bitcoin argue that its energy use is justified as it provides a secure, decentralized, and censorship-resistant monetary system that can operate without the need for intermediaries such as banks. They also argue that Bitcoin mining incentivizes the development of renewable energy sources, as miners seek out cheap and abundant energy sources. Additionally, it is believed that miners put to good use energy that would otherwise be wasted in some parts of the world.

On the other hand, critics argue that Bitcoin's energy consumption is wasteful and unnecessary. They argue that the energy used to mine Bitcoin could be put to better use elsewhere, such as powering homes or businesses. Critics also point out that Bitcoin's energy consumption contributes to climate change, especially when non-renewable sources of energy are used.

So, why does Bitcoin consume so much energy? There are several reasons:

Bitcoin mining requires a lot of computational power to solve complex mathematical problems. This computational power requires a large amount of energy to run the necessary mining hardware, including processors and cooling systems. This ensures the security and integrity of the Bitcoin network by making it difficult for bad actors to manipulate or corrupt the system. A bad actor would need to amass more than half of the total computational power, which costs a lot of electricity and hardware cost, making cheating unprofitable. Thus, Bitcoin's high energy use is also why the Blockchain is very secure.

Bitcoin mining is a highly competitive process, with miners competing to solve mathematical problems before mining a block for bitcoin rewards. This competition drives miners to use increasingly powerful and energy-efficient hardware so as to increase their chances at success. This competitiveness further increases the overall energy consumption of the Bitcoin network. The good part of this is that it helps maintain the decentralization of the Bitcoin protocol. Competition prevents the entire network falling into one entity's control. Thus, as miners strive to outperform each other, the Bitcoin network energy use keeps going up.

With Bitcoin becoming increasingly profitable, new miners are joining the network in their numbers. Several mining pools keep expanding as more participants pool their resources together. This means more computational power enters the network and needs more mining hardware. All this means an increase in energy use. Thus, Bitcoin's energy use keeps increasing and will continue to do so as more miners join the network. This shows that the protocol is accessible to everyone. There's no barriers to entry for those who have the resources.

The increasing value of bitcoin has led to an increase in demand for mining, which in turn has led to an increase in energy consumption. This is because as the value of bitcoin increases, more miners are attracted to the network for the gains (as noted above), which leads to an increase in the overall computing power and energy consumption of the network. The good part about this is that it creates a virtuous cycle in which the value of bitcoin drives innovation and investment in mining hardware.

As the Bitcoin mining difficulty increases, more computational power is required to mine a block which in turn leads to more energy consumption. Mining difficulty is a function of increase in participation in the network. The purpose of the increase is to keep the time it takes to mine one block around 10 minutes so as to control the rate of Bitcoin issuance. Thus, Bitcoin uses so much energy to ensure the currency stays as scarce as possible.

Summary

Bitcoin's high energy consumption is due to the computational power required for mining, the competitive nature of mining, the influx of new miners, the increasing value of Bitcoin, and the mining difficulty. While some may see Bitcoin's energy consumption as a negative, others argue that it is necessary for the security and integrity of the network. Regardless, it's important to be aware of the energy consumption of Bitcoin and consider the environmental impact.

Module D
Bitcoin's Monetary Value

In this module, we will be exploring the monetary value of Bitcoin, one of the most significant and controversial aspects of the cryptocurrency.

We will begin by discussing Bitcoin's role as a form of money and its potential as a medium of exchange. We will also explore the concept of legal tender and how it applies to Bitcoin in different countries.

Next, we will delve into Bitcoin's tokenomics, including its supply, distribution, and inflation. This will provide insight into how Bitcoin's value is determined and how it differs from traditional forms of currency.

We will then examine Bitcoin's role as a store of value, including its volatility and potential for long-term investment. We will also explore the implications of Bitcoin's scarcity and the impact it has on its value.

Finally, we will analyze Bitcoin's market dominance and its position in the global financial system. We will discuss the potential future of Bitcoin as a major player in the world of finance and the challenges it faces in achieving widespread acceptance.

By the end of this module, you will have a solid understanding of Bitcoin's monetary value and its implications for the world of finance. Whether you are interested in investing in Bitcoin or simply want to learn more about this fascinating cryptocurrency, this module will provide more light.

Lesson 21
Bitcoin as Money

This lesson will explore the similarities and differences between Bitcoin and traditional money to help you understand what makes Bitcoin a better form of digital currency.

What is Money?

Let's first define what money is. Money is any item that is widely accepted in exchange for goods or services. It must have value that is widely recognized and trusted by everyone. This value can come from physical commodities, like gold, or digital information, like the numbers in your bank account.

Money is an essential part of our daily lives, and we use it to purchase goods and services, pay bills, save for the future, and more. It is a tool that allows us to exchange one item for another, and it helps us keep track of the value of things.

There are different types of money, such as cash, credit, and digital currencies. Regardless of the form it takes, money must have three main functions: it is a medium of exchange, a unit of account, and a store of value.

Medium of Exchange

The first function of money is as a medium of exchange. This means that money serves as a means of payment, making it easier to trade goods and services without the need for bartering. Bartering involves trading one good or service for another, which can be difficult when the goods or services involved are not easily divisible or are of different values. Money allows for more efficient and flexible exchanges, making it easier for people to obtain what they need and want.

Unit of Account

The second function of money is as a unit of account. This means that money serves as a way to measure the value of goods and services. When we go to the store and see the price of an item, that price is given in a specific currency, such as dollars or euros. The currency acts as a unit of account, allowing us to compare the value of different goods and services.

Store of Value

Money is a store of value. This means that it can be saved and used in the future to purchase goods and services. In order for money to serve as a store of value, it must be trusted and maintain its value over time. If the value of money decreases over time, people may not want to hold onto it, and it will not be an effective store of value.

As noted earlier, money can take many forms, and its value can come from different sources. For example, some money is backed by physical commodities like gold, while other money is based on the trust in a government or financial institution. Bitcoin has gained popularity in recent years, as it offers a decentralized and secure way to exchange value.

Bitcoin as Money

Bitcoin is a form of digital currency, known as a cryptocurrency, that shares some characteristics with traditional money. It's used as a medium of exchange and can be divided into smaller units for little payments, just like traditional money. It's also considered a store of value, meaning that people hold on to it as an investment in the hopes that it will increase in value over time. These are all qualities of money. However, Bitcoin and fiat currencies are not commodities, unlike gold, which is a commodity money.

Since Bitcoin and fiat currencies are not commodities, they do not have intrinsic value. They are only valuable because people believe they are valuable. If people lose interest in them, they become useless. However, Bitcoin is as good as fiat in being considered money because both pass the basic tests for money.

Bitcoin vs Fiat Currencies

One key difference between Bitcoin and traditional money is that Bitcoin is decentralized. This means it is not controlled by any government or central authority. Its transactions are recorded on a public blockchain for increased transparency and security. Furthermore, fiat currencies are legal tender in countries where they are issued. Meanwhile, Bitcoin's legal status as money varies from country to country, with some recognizing it as a legitimate form of currency and others not. This means that the long-term potential of Bitcoin as a form of money is still uncertain.

So, is Bitcoin really money?

Well, it depends on how you define money. While Bitcoin does not have intrinsic value like a physical commodity, it does have value because people trust it and use it as a medium of exchange. In fact, some argue that Bitcoin is a better form of money than traditional money because of its decentralization and transparency.

Let's take a closer look at some of the characteristics that make Bitcoin similar to traditional money or even better:

Durability: Bitcoin is a digital asset that can be stored and transmitted electronically. Unlike physical cash, which shares close semblance with Bitcoin's decentralized nature, Bitcoin is not subject to physical wear and tear, making it more durable in the long run.

Portability: Bitcoin can be easily transferred over the internet, making it highly portable. This means that you can use Bitcoin to make purchases from anywhere in the world without having to worry about physical bulkiness or exchange rates.

Divisibility: Bitcoin can be divided into smaller units, just like traditional digital money. However, Bitcoin is even better than traditional money in this regard because it can be divided into up to 100 million units per Bitcoin.

Acceptability: Bitcoin is accepted by a growing number of merchants and individuals worldwide. Unlike traditional money, Bitcoin can be accepted anywhere in the world without having to worry about exchange rates or currency conversions.

Scarcity: The total supply of Bitcoin is limited to 21 million, giving it scarcity value like traditional money based on physical commodities like gold. Traditional fiat money, on the other hand, can be printed endlessly by governments.

Fungibility: Each unit of Bitcoin is interchangeable with another unit, just like traditional money. However, Bitcoin's fungibility has no national boundaries, unlike traditional money which has to be exchanged for another currency when used in a different country.

No Counterfeit: Bitcoin's blockchain technology makes it difficult to counterfeit, unlike traditional money which is easy to counterfeit.

Store of value: Bitcoin can be used to store value, just like traditional money. Some argue that Bitcoin is a better store of value than traditional money because of its limited supply and decentralization.

Unit of account: Bitcoin can be used as a unit of account. Yes, not yet adopted widely as is the case for most Fiat currencies, but a couple of people are already using bitcoin as a unit of account to determine the value of goods and services.

Summary

Money is any widely accepted item that serves as a medium of exchange, unit of account, and store of value. Bitcoin, a form of digital currency, shares these characteristics. However, Bitcoin and fiat currencies are not commodity money and lack intrinsic value. Bitcoin is decentralized, with transactions recorded on a public blockchain for transparency and security. It has some advantages over traditional money, such as durability, portability, divisibility, acceptability, scarcity, and fungibility. Its long-term potential as a form of money is uncertain due to varying legal status.

Lesson 22
Bitcoin as Legal Tender

In this lesson, we'll explore what a legal tender is and how Bitcoin has become one in some countries.

What is a legal tender?

Legal tender refers to the type of currency that is approved by the government or a particular jurisdiction as a valid means of payment for debts and taxes. This means that if you owe money to someone or the government in that country, you can use the legal tender to pay for it, and no one can legally refuse it.

In the United States, for instance, the national currency is the US Dollar, and it is the legal tender for all debts, public and private. This means that if you owe someone money, you can use US Dollars to pay for it, and they cannot refuse to accept it. This is also true if you owe money to the government, such as taxes. You can use US Dollars to pay for your taxes, and the government must accept it.

It's important to note that not all countries use their national currency as the legal tender. Some countries use foreign currency or a combination of foreign and national currency. For example, Ecuador uses the US Dollar as its legal tender, while Zimbabwe uses a combination of currencies, including the US Dollar, the euro, and the South African rand.

Legal tender status is an important concept because it ensures that people can pay their debts and taxes using a currency that is widely accepted and recognized. It also provides a level of confidence in the currency, as people know that it is backed by the government and is legally recognized as a means of payment.

However, it's important to remember that legal tender does not mean that a currency is the only means of payment. Businesses and individuals can choose to accept other forms of payment, such as credit cards or checks, and they are not legally required to accept legal tender. Additionally, while legal tender must be accepted for debts and taxes, there may be limitations or restrictions on how much can be used, such as in the case of large purchases where the payment is spread out over time.

Making Bitcoin Legal Tender

Currently, only two countries in the world have approved Bitcoin as legal tender - El Salvador and the Central African Republic. On June 9th, 2021, El Salvador's government published in the official gazette the legislation making Bitcoin legal tender within the country. The legislation went into effect on September 7th, 2021. This means that businesses and individuals in El Salvador can use Bitcoin to pay for goods and services, and the government will accept Bitcoin as payment for taxes and other debts.

Similarly, on April 27th, 2022, the Central African Republic announced its adoption of Bitcoin as legal tender. This move aims to improve financial inclusion and help people who do not have access to traditional banking services.

Many other countries have legalized the use of Bitcoin in their jurisdictions without the legal tender status. It is expected that this trend will continue to grow until Bitcoin is eventually a global currency for transaction with international approvals.

But why would a country adopt Bitcoin as legal tender?

There are several reasons why a country might do this. For one, it can provide more financial inclusion to its citizens, especially those who do not have access to traditional banking services. By accepting Bitcoin, individuals can participate in the economy and have access to financial services without having to go through traditional banks.

Additionally, Bitcoin can help reduce the costs associated with traditional financial transactions. Transactions made with Bitcoin are generally faster and cheaper than those made through traditional financial institutions. This can help businesses save money on transaction fees and make it easier for them to conduct business.

However, there are also risks associated with using Bitcoin as legal tender. The value of Bitcoin can be volatile, meaning that its value can fluctuate rapidly. This can make it difficult for individuals and businesses to plan for the future and make financial decisions.

Another risk is that Bitcoin transactions are irreversible. Once a transaction is made, it cannot be reversed or cancelled. This can be a problem if a mistake is made, or if someone is the victim of fraud.

Summary

Bitcoin is a digital currency that can be accepted as a valid form of payment for debts and taxes by the government. Currently, only two countries, El Salvador and the Central African Republic, have approved it as legal tender. It's already legal or not illegal in most countries. While the adoption of Bitcoin as legal tender has its advantages, it also comes with risks that should be carefully considered.

Lesson 23
Bitcoin Tokenomics

Tokenomics is an abbreviation for "token economics," and it refers to the economic and financial aspects of a blockchain-based digital asset, such as Bitcoin. Tokenomics covers the supply, demand, and utility of the token, as well as its market value, issuance, distribution, and governance.

In the case of Bitcoin, tokenomics refers to the economic principles and mechanisms that govern the supply, demand, and utility of the BTC token. Bitcoin's tokenomics is unique in the following ways:

Bitcoin Supply: Limited and Fixed

The supply of Bitcoin is limited to a little less than 21 million units. This means that there can never be more than 21 million Bitcoins in existence. This limit is hard-coded into the Bitcoin core file, ensuring that the value of Bitcoin is determined by market demand rather than being artificially inflated by a central authority. This is protected via Bitcoin's operational game theory via Proof-of-work.

This fixed supply is intended to prevent inflation - a problem facing fiat currencies like the US dollar or Nigerian naira. With a limited supply, Bitcoin's value is protected from inflation caused by an increase in the supply of coins.

Bitcoin Issuance: Mining and Halving

Bitcoin is issued through mining, which involves solving complex mathematical problems to validate transactions on the blockchain. Miners who successfully validate transactions are rewarded with a certain number of Bitcoins from the block, in addition to fees paid for transactions.

In Bitcoin's early days, the reward for mining a block was 50 Bitcoins. This reward was halved after every 210,000 blocks, which takes about 4 years. So far, this event known as Bitcoin halving has occurred three times, cutting the block reward to 25, then 12.5, and finally to its current quantity of 6.25 BTC per block. The next halving will occur in 2024, further reducing the block reward to 3.125 BTC per block. This will continue to happen in 4-year intervals until all Bitcoin supply has been issued.

The purpose of halving is to control the rate at which new Bitcoins are released into circulation and maintain the fixed supply of 21 million units in the long run. The last Bitcoin will be mined around the year 2140, dependent on the time it takes to mine one Bitcoin block which can fluctuate as more miners leave or join the network.

Bitcoin Demand and Liquidity

The demand for Bitcoin is driven by its perceived value as a decentralized, secure, and globally recognized digital asset. Its liquidity is determined by the number of exchanges and merchants that accept it as a form of payment, as well as the number of individuals and institutions holding it as a store of value.

Bitcoin's popularity has grown rapidly over the years, making it the most popular and in-demand cryptocurrency. It's constantly traded on exchanges and peer-to-peer platforms, making it highly liquid and accessible.

Bitcoin Utility: Store of Value, Medium of Exchange, Unit of Account

Bitcoin can be used as a store of value, a medium of exchange, and a unit of account. Its utility is determined by its perceived value and the number of individuals and institutions that recognize and accept it as a valid form of payment.

In some countries like El Salvador and Central African Republic, Bitcoin is already recognized as a legal tender. This means individuals can use it to pay for goods and services in these countries. In many other countries, the use of Bitcoin is legal, prompting involvement by institutions like banks, hedge funds and others holding Bitcoin. So many businesses around the world accept it for payments.

Bitcoin Divisibility

One whole Bitcoin can be divided into 100 million smaller units called Satoshi. This makes it highly flexible for use locally for very small transactions as little as a cent.

Summary

Tokenomics is the economics and finance of digital assets, like Bitcoin. Bitcoin's tokenomics include a limited and fixed supply, issuance control through mining and halving, high demand and liquidity, and its use as a store of value, medium of exchange, and unit of account. Bitcoin's divisibility allows for small transactions, and its popularity has grown making it accessible and widely accepted as a valid form of payment.

Lesson 24
Bitcoin as Store of Value

Bitcoin is a digital currency that has been around for over a decade. It has survived so far because many people interested in it see it as a good store of value. But what does that mean? Why is Bitcoin a good store of value?

A store of value is something that holds, retains or improves its value over time. In other words, if you have something that is valuable today, you want to be able to hold onto that value for a long time. Think of it as a way to keep your money safe and secure, without worrying about it losing its purchasing power. Traditionally, people have used physical assets such as gold, silver, and real estate as stores of value. However, Bitcoin has emerged as a new contender in the digital age. Surprisingly, it has also outperformed all its competitors since its launch.

What Makes Bitcoin a Store of Value?

To be a good store of value, something must have three main features:

Durability: The item must be able to last a long time without deteriorating in value. For example, gold is a durable item because it does not corrode or rust over time.

Acceptability: The item must be widely accepted as having value. If no one accepts it, it's not very useful. For example, paper money is widely accepted as having value because people trust that they can use it to buy goods and services.

Easily Interchangeable: The item must be easily exchangeable for other items of value. If something is not easily exchangeable, its usefulness as a store of value is compromised. For example, if you have a rare painting that is worth a lot of money, but no one is willing to buy it, it's not very useful.

Bitcoin meets all of these criteria, which is why it is considered a good store of value.

So, what makes Bitcoin such a good store of value? Let's take a closer look at some of the key factors I have outlined below:

Scarcity

One of the primary reasons why Bitcoin is considered a good store of value is because of its scarcity. Unlike fiat currencies, which can be printed by central banks at will, there is a limited supply of Bitcoin. In fact, there will never be more than 21 million Bitcoins in existence. This creates a sense of value and ensures that the coin cannot be debased through inflation. Contrary to the way fiat responds to inflation, Bitcoin increases it's value as demand and adoption for it increases.

Decentralization

Another important feature of Bitcoin is its decentralized nature. Unlike traditional currencies, which are controlled by governments or financial institutions, Bitcoin is not owned or controlled by any central authority. This means that users have more autonomy and freedom in terms of how they use and store the coin. There's no fear of single points of failure that may lead to loss of funds.

Immutable

Bitcoin's blockchain technology ensures that all transactions are recorded immutably, meaning that they cannot be altered or deleted. This adds to the coin's security and trustworthiness, as there is a permanent record of every transaction that has ever taken place. This gives users the confidence that what they store cannot be counterfeited nor destroyed.

Liquidity

Bitcoin is widely accepted and traded on various exchanges and platforms, making it easy to buy and sell. This means the currency is easy to exchange at any given time. With increasing adoption and mainstream participation, storing ones wealth in Bitcoin is a potentially beneficial initiative in the long term.

Recognition

Over the past few years, Bitcoin has gained recognition as a legitimate store of value by many individuals and institutions. This includes major companies, investors, and even some governments. This recognition adds to the coin's credibility and perceived value. Some traditional financial institutions already have Bitcoin in their balance sheet.

Transparency

Another benefit of Bitcoin's blockchain technology is that it allows for transparency in all transactions. This makes it easier for users to track their coins and ensure their security. Thus, Bitcoin is a store of value whose storage system can be monitored remotely without requiring any expensive hardware or security. You can always check that your funds are intact by looking at the publicly viewable Bitcoin Blockchain.

Security

Bitcoin is designed to be highly secure, with advanced encryption and other measures in place to protect against hacking and fraud. While no system is completely foolproof, Bitcoin's security features make it much safer than traditional forms of currency. Thus, it's ideal for securing value over a long period of time without a significant risk of losing the saved value.

Borderless

Perhaps one of the most important features of Bitcoin is its borderless nature. Because it is a digital currency, Bitcoin can be sent and received anywhere in the world, instantly and at low cost. This enables people to store value outside of their own country's borders, protecting it from local political and economic instability.

Deflationary

Finally, Bitcoin has a predefined monetary policy, which makes it inherently deflationary. As the global supply of Bitcoin will never exceed 21 million coins, the purchasing power of each coin will increase as the number of people who want to own Bitcoin increases. This makes it a good hedge against inflation. So, being deflationary is a reference to how the value of Bitcoin increases with passage of time instead of depreciating.

Summary

Bitcoin is considered a good store of value because it is durable, widely accepted, easily interchangeable, scarce, decentralized, immutable, liquid, recognized, transparent, secure, borderless, and deflationary. However, it's important to note that Bitcoin is also highly volatile, which means that its short-term value can fluctuate wildly. As with any investment, it's important to do your research and understand the risks before investing in Bitcoin.

While Bitcoin has been recognized by many as a store of value, there are still concerns about its long-term stability and potential regulation by governments. Additionally, as with any investment, the value of Bitcoin can be affected by various factors such as market demand, technological developments, and overall economic conditions.

Lesson 25
Implications of Bitcoin Scarcity

Scarcity is a fundamental concept in economics that refers to the limited availability of resources in relation to the unlimited wants and needs of individuals and societies. In the world of cryptocurrency, Bitcoin is a prime example of a scarce resource due to its limited supply of 21 million Bitcoins - the maximum that will ever exist.

What does this mean for Bitcoin and its users? Let's explore the implications of Bitcoin's scarcity for users below:

Value Appreciation

Bitcoin is a type of digital currency that people can use to buy things or invest in. As more people start using Bitcoin, the demand for it increases. However, the amount of Bitcoin available is limited, so the price of Bitcoin goes up as more people want to buy it. This makes Bitcoin similar to gold or other precious metals, which are also valuable because they are scarce. Many people think that Bitcoin is a good investment because it has a limited supply, so its value may continue to increase over time. Think of it like a rare baseball card - the more people want it, the more valuable it becomes!

Inflation Protection

Inflation happens when there's too much money circulating in the economy, and people start competing with each other to buy the same goods and services. This leads to higher prices, and the value of money decreases. Bitcoin is different because there's a limited amount of it that will ever exist (only 21 million bitcoins). Additionally, as more people start using it, it becomes harder to create new bitcoins (mining difficulty increases). This makes Bitcoin a good way to protect yourself against inflation, because even if the value of regular money decreases, the value of Bitcoin could stay the same or even go up over time. So, Bitcoin can help you protect your savings from losing value due to inflation.

Network Effect

The network effect of Bitcoin is similar to how social media works. When you join a social media platform, it becomes more useful to you as more people join it. With Bitcoin, as more people use it, the more valuable it becomes. The value of Bitcoin is not just in its technology, but also in the number of people who are willing to use it. This is why the more people that adopt Bitcoin, the more valuable it becomes in the long run.

As Bitcoin is scarce and limited in supply, it creates a situation where demand can be much higher than supply, resulting in an increase in price. This scarcity is what makes Bitcoin unique, and the network effect is what gives it value.

Limited Downside Risk

Bitcoin is a digital currency that has a fixed amount, which means that no more can be created beyond its 21 million cap. This means that it is immune to inflation and government manipulation, which is a common problem with traditional currencies that can be printed or manipulated by governments. This fixed amount of Bitcoin also means that no single person or group can control or manipulate its value.

Think of it like a limited edition toy that only has a certain number of pieces made. If you own one of those toys, it becomes more valuable because there are only a limited number of them in existence. Similarly, because there are only a limited number of Bitcoins, their value can increase as demand for them grows. This limited downside risk is what makes Bitcoin an attractive option for investors who want to protect their money from inflation and government interference.

Improved Desirability

Bitcoin's scarcity creates a sense of rarity and exclusivity around the asset, making it desirable for those who can acquire it. Because of this, people who own Bitcoin feel special and unique because they have something that not everyone else can get. As more people become interested in Bitcoin, the demand for it goes up. Since there is only a limited supply, the price of Bitcoin is likely to keep going up too. This is good news for people who bought Bitcoin early on because they were able to get it for a far lower price.

Summary

Bitcoin's scarcity is a crucial element that impacts its value and potential as an investment asset. As adoption and demand for Bitcoin increase, the scarcity puts pressure on the limited supply, leading to an appreciation in value. Bitcoin's scarcity also offers protection against inflation, a network effect that drives speculation, limited downside risk, and increased desirability for early adopters and investors.

Lesson 26
Bitcoin Market Dominance

If you're new to the world of cryptocurrencies, you might have heard the term "Bitcoin Dominance" being thrown around. In simple terms, Bitcoin Dominance is a metric that measures the relative dominance of Bitcoin in the cryptocurrency market. It compares Bitcoin's value to the overall value of all the other Cryptocurrencies, called altcoins.

To understand Bitcoin Dominance, let's use an analogy. Imagine all the cryptocurrencies in existence as a big, juicy apple. The size of this apple represents the total market capitalization of all cryptocurrencies combined, including Bitcoin's. Then you decide to take out one big slice from the full apple. This slice is equivalent to Bitcoin's market capitalization. Bitcoin Dominance is the ratio of the slice of the apple that represents the market capitalization of Bitcoin compared to the total market capitalization of all cryptocurrencies, that is, without any slice cut off.

Market capitalization is a way to measure the size and value of a cryptocurrency. It is calculated by multiplying the price of a cryptocurrency by the total number of coins or tokens in circulation. For example, if a cryptocurrency has a price of $100 and there are 10 million coins in circulation, its market capitalization would be $1 billion (100 x 10,000,000).

To calculate Bitcoin Dominance, you take the market capitalization of Bitcoin and divide it by the total market capitalization of all cryptocurrencies put together (including that of Bitcoin). You then multiply this number by 100% to get the percentage of Bitcoin Dominance.

Bitcoin Dominance = (Market Capitalization of Bitcoin / Total Market Capitalization of All Cryptocurrencies) x 100%

For example, if the market capitalization of Bitcoin is $1 trillion and the total market capitalization of all cryptocurrencies is $2 trillion, then Bitcoin Dominance would be 50% (1 trillion / 2 trillion x 100%).

Why is Bitcoin Dominance important?

Well, it provides a snapshot of how dominant Bitcoin is in the cryptocurrency market. A high Bitcoin Dominance means that Bitcoin is the dominant player in the market, while a low Bitcoin Dominance means that other cryptocurrencies are gaining ground.

It's worth noting that Bitcoin Dominance can change over time. In the early days of cryptocurrencies, Bitcoin Dominance was close to 100%, meaning that Bitcoin was the only cryptocurrency worth talking about. However, as more cryptocurrencies were created, Bitcoin Dominance started to decline. At the time of writing, Bitcoin Dominance is around 45%, which means that other cryptocurrencies have a significant presence in the market.

What does this mean for investors?

If you're bullish on Bitcoin, a high Bitcoin Dominance might indicate that Bitcoin is a good investment opportunity. On the other hand, if you're interested in other cryptocurrencies, a low Bitcoin Dominance might suggest that there are more opportunities in the market outside of Bitcoin.

It's important to note that Bitcoin Dominance is just one metric to consider when investing in cryptocurrencies. Other factors to consider include the technology behind the cryptocurrency, the team behind the project, and the potential use cases for the cryptocurrency. This course does not focus on providing investment ideas. Consult previous lessons in this book to understand more about factors listed above.

Summary

Bitcoin Dominance is a metric that represents the percentage of the total market capitalization of cryptocurrencies that is attributed to Bitcoin. It provides a measure of how dominant Bitcoin is in comparison to other cryptocurrencies in the market, calculated in terms of financial value. The formula for Bitcoin Dominance is (Market Capitalization of Bitcoin / Total Market Capitalization of All Cryptocurrencies) x 100%.

Module E
Holding and Trading Bitcoin

This module is specifically designed to help you understand the practical aspects of using and investing in Bitcoin, including how to purchase Bitcoin, store it securely, send and receive it, and even trade it on various exchanges.

You will learn the various ways to purchase Bitcoin, from traditional exchanges to peer-to-peer platforms. You will learn how to choose the right platform for you, create account and make your first bitcoin purchase.

You will learn how to store your Bitcoin safely and securely. We will explore the various types of wallets, including software wallets and hardware wallets. You will also learn how to use these wallets to ensure the safety of your Bitcoin.

You will learn how to send Bitcoin from your wallet, and how to receive Bitcoin from others.

You will equally learn how to trade Bitcoin on various exchanges profitably. We will explore the different types of exchanges and trading platforms, including centralized and decentralized exchanges.

Finally, you will also learn about the risks associated with holding and trading Bitcoin, including volatility, security risks, and scams. You will learn how to mitigate these risks and how to protect yourself from potential losses.

By the end of this module, you will have a solid understanding of how to hold and trade Bitcoin safely and effectively. You will be equipped with the practical knowledge you need to start investing in Bitcoin with confidence if you choose.

Lesson 27
How to Buy Bitcoin

If you're interested in buying Bitcoin, there are several ways to do so. However, before purchasing any bitcoin, it's essential to know the risks involved, do your research, and consider factors like fees, regulations, security, and the reputation of the platform you want to use.

Let's take a look at some of the most common methods used to purchase Bitcoin today.

Centralized Exchanges

Centralized exchanges, also known as CEX, are online platforms that allow users to buy and sell cryptocurrencies. CEXs enable users to buy Bitcoin with their local currency. This largely depends on their country of residence and whether Bitcoin-related financial transactions are acceptable by financial institutions. For example, while people in the US can buy Bitcoin directly from their bank accounts or with credit/debit cards, others in countries like Nigeria may not due to restrictions by local financial regulators on Bitcoin-related transactions.

Some examples of CEXs include Coinbase, Binance, Bybit, MEXC GLOBAL, and Kraken.

Buying Bitcoin from a CEX comes with advantages such as user convenience (their interfaces are relatively easy to understand) and wide availability (they are open round the clock, unless there's a downtime). Users can purchase Bitcoin using bank transfer or credit card. However, one disadvantage of using CEX is that they charge high fees, may have more complex user interfaces than other options, and require users to verify their identity. KYC verification removes the key Bitcoin feature of anonymity and identity privacy.

To buy Bitcoin from an exchange, you need to download and register via their mobile app or website. It's important to note that there are many counterfeit apps and websites imitating popular exchanges. The best way to ascertain the right apps and websites is to locate the verified social media accounts of these exchanges and follow the information in their bio to avoid scammers.

Peer-to-peer Marketplaces

Peer-to-peer (P2P) marketplaces enable individuals to buy Bitcoin directly from each other using digital fiat or cash transfer. P2P platforms act as brokers through the use of escrow services. An escrow is similar to smart contracts but is not blockchain-based. It enables a trustless transaction to take place.

Two individuals can initiate a transaction with each other using the escrow. The seller locks their Bitcoin in the escrow while awaiting payment from the buyer directly to their bank account. Once the buyer has made payment, they notify the escrow, which notifies the seller, who then checks that they have received the payment. Once payment is confirmed, the seller notifies the escrow to release the Bitcoin to the buyer.

Some examples of P2P marketplaces include LocalBitcoins (now defunct) and Paxful. These platforms allow users to buy and sell Bitcoin directly with one another, rather than through a centralized exchange. Today, many popular exchanges also have P2P marketplaces on their platforms to make it easy for new users who can't connect their bank accounts or credit card to still deposit via P2P. Old users who need cash can still use the platform to sell their Bitcoin for cash.

Advantages of P2P marketplaces include potentially lower fees and the ability to purchase using cash or other methods not supported by traditional exchanges. However, disadvantages can include a lack of regulatory oversight and the potential for fraud. For example, a buyer who fails to notify the escrow after paying might lose their money if the time set by the escrow expires. Similarly, a seller who releases their Bitcoin without confirming payment might not get paid.

In the event of a dispute, the buyer can raise a dispute with the escrow, requiring them to show proof of payment, while the seller will need to submit their bank statement to the escrow for verification.

Bitcoin ATMs

Bitcoin ATMs are physical machines that allow you to buy Bitcoin using cash or a debit card. Some machines also allow you to sell your Bitcoin for cash. To buy Bitcoin from a Bitcoin ATM, you'll need cash or money in your bank account or credit card. You'll also need a non-custodial wallet address where you want to receive your Bitcoin after purchase.

The process of buying Bitcoin at a Bitcoin ATM is relatively simple. First, locate a Bitcoin ATM near you. Then, log in to your account if required. You may be prompted to enter your Bitcoin address or let the ATM scan your Bitcoin receiving QR code. Finally, you'll need to either insert your credit card to pay digitally or insert cash via the slot provided.

While Bitcoin ATMs offer a quick and easy way to buy Bitcoin with cash, they often come with higher fees than other methods.

Over-the-Counter (OTC) Marketplaces

OTC marketplaces are similar to peer-to-peer platforms, but they use centralized exchanges that are set up for institutional bodies or people with large funds seeking to buy Bitcoin. These platforms act as brokers to connect buyers seeking large amounts of Bitcoin with sellers who have matching trades.

OTC traders or platforms help you buy large amounts of Bitcoin directly from another person, rather than placing orders through an exchange. This method is well-suited for large transactions involving institutional investors, but it can be less transparent, less regulated, and less convenient than other methods. There are often minimum amounts that can be traded via OTC, often in the five or six figures range.

Summary

This lesson provides an overview of the different methods available for buying Bitcoin, with the reminder that all methods carry their own risks. The options discussed include Bitcoin ATMs, Over-the-Counter (OTC) marketplaces, and peer-to-peer marketplaces, each with its own advantages and disadvantages. For those looking to buy Bitcoin using cash, Bitcoin ATMs and peer-to-peer marketplaces such as Paxful are good options. For larger transactions, OTC marketplaces provide a more secure and confidential option. Exchanges such as Binance, Coinbase, Bybit, or MEXC offer the convenience of buying Bitcoin directly using bank accounts or credit cards.

Lesson 28
Storing Bitcoin Securely

If you are new to the world of Bitcoin, you may be wondering how you can best keep your investment secure. The rule of thumb is that you should hold your Bitcoin in a self-custody wallet. This means that you are responsible for securing your private keys that give you access to your coins on the blockchain. In this lesson, I will explain how to store your Bitcoin safely and the different types of wallets you can use.

What does it mean to Securely Store Bitcoin?

Bitcoin is a digital currency that is not stored in a bank account or physical wallet like traditional currencies. Instead, it is stored on a decentralized digital ledger called the Bitcoin blockchain.

When you buy Bitcoin, you are given a private key, which is like a password that gives you access to your Bitcoin on the blockchain. This key is important to keep secure because if someone else gets access to it, they can steal your Bitcoin. If you buy Bitcoin on a centralized exchange, the exchange holds the private key on your behalf but gives you access to your funds.

To store Bitcoin securely on your own, you need a Bitcoin wallet. This is a software program that you install on your computer or mobile device. The wallet generates a public key using a private key that grants you access to a space on the Bitcoin Blockchain where your Bitcoin can be stored. The public key is a code that other people can use to send Bitcoin to you. Your Bitcoin address is hashed from the public key, and it's safe to share with others because it only allows people to send Bitcoin to you, not access or steal it.

Your Bitcoin wallet private key is a code that only you should know. This key gives you access to your Bitcoin on the blockchain, so you can spend it or transfer it to someone else. It's crucial to keep your private key secure because if someone else gets access to it, they can steal your Bitcoin.

There are different types of Bitcoin wallets. Generally, these include hardware wallets, software wallets, and online/web wallets. Hardware wallets are the most secure because they are physical devices that store your private key offline, away from the internet. Software wallets are installed on your computer or mobile device, and online wallets are web-based services that store your private key on their servers.

There are other classifications of wallet types. This is explained in the section below, under the 'Best Wallet to use.'

Overall, securely storing Bitcoin means keeping your private key safe and using a reputable Bitcoin wallet to access your Bitcoin on the blockchain. By doing this, you have complete control over your funds and can ensure that no one else can access or steal your Bitcoin.

Best Wallet to use

If you want to store your Bitcoin securely, you need to understand the difference between two types of wallets: self-custody and custodied wallets.

Self-custody wallets give you complete control over your private keys and you're responsible for keeping your Bitcoin safe. You can store your Bitcoin on a personal device like a computer or mobile phone. This method is the most secure way to store Bitcoin since only you have access to your private keys.

Custodied wallets, on the other hand, are when you store your Bitcoin with a third-party service like an exchange or a custody provider. These services are responsible for the security of your Bitcoin, which makes them less secure than self-custody wallets. However, custodied wallets may be more convenient since you don't have to worry about the technical details of keeping your Bitcoin secure.

You should also be aware of the difference between hot wallets and cold wallets. Hot wallets are Bitcoin wallets that have immediate access to the internet, such as mobile software wallets, web wallets, and desktop wallets. These wallets are more vulnerable to hacking or theft since they are connected to the internet. Cold wallets, on the other hand, are Bitcoin wallets that are not connected to the internet, such as hardware wallets and paper wallets. These wallets are the most secure since they are not accessible to hackers.

The best wallet to use for storing your Bitcoin depends on your needs and level of technical expertise. If you want maximum security and control over your Bitcoin, a self-custody cold wallet is the best option. However, if you prefer convenience over security, a custodied hot wallet might be more suitable.

How to Securely Store your Bitcoin

If you have Bitcoin and want to store it securely, there are some steps you can follow to keep your funds safe:

Use self-custody wallets: Self-custody wallets give you complete control over your Bitcoin, meaning you are the only one who has access to your funds. This reduces the risk of losing your funds in unforeseen circumstances beyond your control, such as exchange hacks or failures.

Use a hardware wallet: Hardware wallets are small physical devices that store your private keys offline. This means that they are not connected to the internet, making them much less vulnerable to hacking attempts. Hardware wallets are considered one of the most secure ways to store Bitcoin.

Backup your private keys: Private keys are what allow you to access your Bitcoin. It is important to keep them safe and back them up in case of loss or damage. You can do this by writing them down on a piece of paper and storing them in a safe place, such as a safe or a safety deposit box.

Keep your wallet software up-to-date: Wallet software is regularly updated with new security features to protect your funds from potential threats. Make sure to always use the latest version of your wallet software to stay protected.

Use two-factor authentication: Two-factor authentication (2FA) adds an extra layer of security to your Bitcoin wallet by requiring a code or confirmation in addition to your password. This is particularly important if you are using a web wallet or a custodial wallet.

Remember, storing your Bitcoin securely is crucial to protecting your investment. It is recommended to hold your Bitcoin in a self-custody wallet and use a hardware wallet for maximum security. Always backup your private keys and keep them in a safe place. By following these steps, you can feel more confident in the safety of your Bitcoin.

Summary

Bitcoin is stored on a digital ledger called the blockchain. When you buy Bitcoin, you get a private key that gives you access to it. You need to keep this key secure, or someone else could steal your Bitcoin. To do this, you need a Bitcoin wallet, which is a program that generates a public key you can share to receive Bitcoin and holds the private key you should keep safe.

There are different types of Bitcoin wallets, including hardware, software or online, and self-custody or custodied wallets. Self-custody wallets are the most secure, but custodied wallets may be more convenient. Hot wallets, which are connected to the internet, are more vulnerable to hacking than cold wallets, which are offline. If you want maximum security and control over your Bitcoin, a self-custody cold wallet is the best option.

To store Bitcoin securely, use self-custody wallets, a hardware wallet, backup your private keys, keep your wallet software up-to-date, and use two-factor authentication.

Lesson 29
Sending and Receiving Bitcoin

Are you interested in sending or receiving Bitcoin securely? It's crucial to do so to prevent losing your coins to unknown individuals or scammers. This lesson will explain what you need to do while sending or receiving Bitcoin transactions to make them safe.

Safety Tips Before Starting a Bitcoin Transaction

Here are some important things to keep in mind before sending or receiving Bitcoin:

1. Protect your Bitcoin wallet with a strong and unique password or passcode. Don't use the same password for any other accounts, and only share it with people you trust. However, keep in mind that this password only protects your wallet from physical tampering, and not from someone with access to your private key or seed phrase.

2. If you are using a centralized exchange Bitcoin wallet, it's safer to enable two-factor authentication (2FA) for your wallet. This adds an extra layer of security by requiring a code from your phone, in addition to your password. It's also important to ensure that the phone you receive the code on is different from the one that has the

Bitcoin wallet. This will prevent someone from stealing your funds without your knowledge.

3. Always double-check the address before sending any Bitcoin funds. Verify that the address is correct and belongs to someone you trust. If someone sends you their address, ask them to confirm it multiple times to make sure they're not sending you the wrong one. Also, ensure that the counterparty double-checks the address you sent to them. Be careful not to fall for malware that can replace a scammer's address with the one you copied.

4. Install anti-virus software on your computer and mobile devices to keep them secure. Keep your apps up-to-date and regularly scan your devices for malware, especially if you're always browsing the internet over public Wi-Fi.

5. Avoid using public Wi-Fi networks when handling Bitcoin transactions, such as those in your school or public places. Use your private phone internet instead. Public Wi-Fi networks may not be secure, and hackers may be watching for Bitcoin-related transactions.

6. Never share your private keys or seed phrases with anyone. These codes give access to your Bitcoin, and sharing them with others can put your funds at risk. Even if the support or developers of a wallet you use ask for your seed phrase, don't give it to them. They will never ask for it. If anyone asks

for it, they are scammers, and you should not trust them.

Remember, Bitcoin transactions can be irreversible and are not protected by any government. Therefore, it's essential to take the necessary precautions to protect your funds from potential theft or loss. By following these tips, you can ensure that your Bitcoin transactions are safe and secure.

How to SEND Bitcoin

Sending Bitcoin requires a few steps to ensure the process goes smoothly. To start, you will need a non-custodial wallet, which is a secure place where you can store and control your Bitcoin. Once you have set up a wallet, and funded it, you can send Bitcoin to anyone with a Bitcoin wallet address.

To send Bitcoin, follow these steps:

1. Open your Bitcoin wallet and go to the "Send,""Withdraw," or "Transfer" section. This is where you will initiate the transaction.

2. Copy the recipient's Bitcoin address. It is important to double-check the address to ensure that it is correct. If you send Bitcoin to the wrong address, you may not be able to recover the funds.

3. Enter the amount of Bitcoin you want to send. Be careful not to input the wrong amount, especially if you are sending Bitcoin to someone you don't

know well. If you send too much Bitcoin to someone, you may not be able to get a refund.

4. Your wallet will automatically calculate the transaction fee. This fee goes to network miners who process the transaction.

5. Review the transaction details to make sure they are correct. If everything looks good, click "Send" or "Confirm" to initiate the transaction.

6. The transaction will be broadcast to the network, where it will be verified and processed by nodes on the network. This process may take a few minutes, during which time the funds will be in a "pending" state.

7. Once the transaction is confirmed, the recipient will be able to see the funds in their wallet. It typically takes about an hour for a transaction to receive six confirmations, which is considered secure.

Remember to always be careful when sending Bitcoin. Double-check the recipient's address and the amount you want to send. And if you have any doubts, don't hesitate to reach out to someone with experience in Bitcoin transactions.

How to RECEIVE bitcoin

Receiving bitcoin is easier than sending it. Here's how you can receive bitcoin:

1. First, open your Bitcoin wallet. If you don't have one, you can download it from the app store or play store.
2. Next, click on the "Receive" or "Deposit" section. This is where you will get your Bitcoin address.
3. Copy your Bitcoin address carefully. It is a long string of letters and numbers. You can also use the QR code if the sender is nearby.
4. Double-check the address to ensure it's correct. Then, share it with the sender. It's important to share the correct address; otherwise, you might not receive the bitcoin.
5. Remind the sender to double-check the address before they send the bitcoin.
6. The sender will enter your address as the recipient when they initiate the transaction. This is similar to how you send bitcoin out.
7. Once the sender confirms the transaction, it will be sent to the network and processed by nodes. This process may take a few minutes.
8. After the transaction is confirmed, the funds will be credited to your wallet. You can check your "Transactions" or "History" section to see the new transaction. The bitcoin display will show the value in your currency of choice.

Remember that bitcoin transactions cannot be reversed once they are confirmed on the blockchain. So, it's crucial to double-check the address and the amount before sending any bitcoin. And if you're receiving bitcoin, make sure you cross-check the address before forwarding it to the sender.

Summary

If you want to send or receive Bitcoin securely, there are some important things to keep in mind. First, use a strong password and enable two-factor authentication if possible. Always double-check the address you're sending Bitcoin to, and don't use public Wi-Fi. Never share your private keys or seed phrases with anyone. When sending Bitcoin, copy the recipient's address, input the amount you want to send, and confirm the transaction details before clicking send. To receive Bitcoin, open your wallet and get your Bitcoin address, double-check it, and share it with the sender. Remember that Bitcoin transactions can't be reversed, so always be careful.

Lesson 30
Trading Bitcoin

Do you want to start trading Bitcoin but don't know where to start? Well, I've got you covered! In this lesson, I will guide you through the basics of trading Bitcoin profitably.

First things first, to be able to actively trade Bitcoin, you need to create an account on a popular cryptocurrency exchange such as MEXC, Binance, or Bybit. However, using a popular exchange is not a guarantee of making profits or safety. It is essential to note that the safest way to hold Bitcoin is via a non-custodial wallet. Thus, exchanges are only good for active traders.

Secondly, after creating an account on a cryptocurrency exchange, you need to buy your first Bitcoin. But before you buy your first Bitcoin, you must know that the Bitcoin market is very volatile, and it is very risky to put more money than you can willingly lose. Therefore, never use money you need urgently to buy Bitcoin, whether you're trading or investing.

Thirdly, you need to understand the difference between investing and trading Bitcoin. Investing in Bitcoin refers to simply buying Bitcoin and holding it in a non-custodial wallet for a very long time, hoping for the price to appreciate. Trading Bitcoin involves buying Bitcoin at a low price and selling at a higher price to make a quick profit. Trading can occur in a minute, hour, day, week, or a few months, while investing usually takes several years or as long as your profit target for the investment is hit.

Lastly, you should never give money to anyone to trade or invest in Bitcoin on your behalf. Only compromise is if you want to pool funds with a Bitcoin trust fund like microstrategy, a Bitcoin investment company. Do not invest with any trust fund that trades with the Bitcoin they're holding for investors. The goal is to buy Bitcoin every time it dips and hold for ten years in a self-custody cold wallet. All participants will be able to monitor this address 24/7.

Tips for trading Bitcoin Profitably

To make your Bitcoin trading profitable, below are the things to do. Note that this is only a lesson and should not be taken as a financial advise.

A. Conduct thorough research and analyze the Bitcoin market before making any trades on any exchange. You can check the liquidity in an exchange and see if there's any history of market manipulation. Read reviews by other users. Bitcoin itself is a great crypto instrument to trade. All the information you need about it can be seen on a coin data aggregator like coinmarketcap.

B.   Develop a solid trading strategy and stick to it, avoiding impulsive or emotional decisions. By trading strategy, we mean deciding whether you want to trade in a short term, medium term, or long term. It also means deciding which Bitcoin market to trade on. We have Spot (where you simply buy low and sell high) and Futures (where you simply predict Bitcoin price movements). Spot is safer for beginners as there is fewer ways you can lose all your money. Futures is riskier since you can lose all your capital unexpectedly. Therefore, choose your time frame, your type of trade, and finally, how much you want to use per trade every time. Stick to this strategy and don't change it simply because you're excited or afraid.

C.   Keep a close eye on market news and events that could potentially impact the price of Bitcoin. The latest news most times affects the market. The reason is that, contrary to advice, most people trade Bitcoin based on their emotions. For instance, if the news is bad about Bitcoin, many people start selling off to avoid getting stuck. When this happens, prices fall. If the news is good, everyone wants to buy Bitcoin so they don't miss out. When this happens, prices rise. When you pay attention to news, you will learn that the best time to buy Bitcoin is during dips or corrections. Corrections happen when the price of Bitcoin drops by a significant percentage, and it is a common occurrence in the crypto market.

Corrections are normal and are not necessarily an indication of a bearish trend. Instead, they provide a buying opportunity for traders and investors to buy Bitcoin at a lower price and wait for the price to increase again.

D. Diversify Your Portfolio - One of the most important things to remember when trading bitcoin is to diversify your portfolio. Don't put all your eggs in one basket by investing solely in bitcoin. Instead, invest in a variety of assets such as stocks, real estate, agriculture, or travel and tourism. This will help you spread your risks and minimize your losses if one asset doesn't perform well.

E. Use Proper Risk Management Strategies - Another crucial aspect of trading bitcoin profitably is to use proper risk management strategies. For instance, you should always use stop-loss orders to limit potential losses in case the market takes an unexpected turn. You should also take profit every time you see it. As a trader or investor, you can never go wrong with taking profit. Your risk-to-reward ratio should not be less than 1:2, which means that if you're comfortable with losing 10% of your bitcoin trade, you should always set your profit twice that (20%) using the take-profit function on your favorite exchange. Also, whenever your trade goes into profit, immediately move your stop loss into profit. For example, if your bitcoin trade is at +15% profit, move your stop loss from -10% to

+10%. This way, if the price retraces before reaching your 20% profit target, you can walk away with a 10% profit. If you want to learn more about risk management strategies, you can contact the author for a one on one guideline.

F.   Keep Track of Your Trades - To improve your trading strategy, you should keep track of your trades and review them regularly. Never ignore an open trade, but instead watch it from time to time, depending on your trading strategy, to make all necessary adjustments. Additionally, record your trade activities in a journal. This will help you understand if a strategy works for you or if you need to make some changes to improve.

G.   Stay Up-to-Date on Technical Analysis and Charting Techniques - To make informed trading decisions, you should stay up-to-date on technical analysis and charting techniques. You should learn how to read bitcoin trading charts using basic technical indicators like Moving Average (Exponential), Relative Strength Index, and Bollinger Band. You can equally learn to read candlestick chart patterns. You can learn about these things through free bitcoin educational documents available online. However, if you want to learn more in-depth, you can contact me for a guide.

H.   Use a Reliable and Secure Platform for Trading and Storing Bitcoin - The exchange you use for trading bitcoin matters. Use popular exchanges and, if you're not actively trading, never keep your funds on these exchanges. None of the exchanges I mentioned in this book pay me directly for promotion, and I am not responsible for any losses you may incur by using anyone of them. You cannot hold anybody accountable if you trade on an exchange that goes bankrupt or fails due to bank run or mismanagement of user funds. Always do your own research and understanding the risks before investing or trading on any platform.

I.   Keep Your Emotions in Check - To be a successful trader, you must gain mastery of your emotions. Avoid FOMO (fear of missing out) or FUD (fear, uncertainty, and doubt) when trading. Remember that trading based on news only can be misleading. Other indicators should be examined before every decision is finalized. Being in control means you know the right time to buy or sell, and you don't get too excited or afraid of anything. Just be aware of the risk and use only what you have no worries losing or waiting for. This will help you control your emotions.

J.   It's essential to continuously educate yourself and stay informed about the latest developments in the Bitcoin market. Knowledge is power, and as a consumer of knowledge, you should never stop trying to learn more about Bitcoin and trading strategies that work. If you stop learning, you might fall behind on current trends, and that could lead to losses.

To keep yourself up-to-date with the latest trends, follow my CryptoStarterTV social media platforms, including on YouTube, Facebook, Instagram, Twitter, and Telegram. Our website is also a great place to learn more about Bitcoin trading in-depth.

Summary

This lesson only shared relevant tips on how to trade Bitcoin profitably. The first step is to create an account on a popular cryptocurrency exchange. The second step is to understand the difference between investing and trading Bitcoin. Investing involves holding Bitcoin for a long time, while trading involves buying and selling Bitcoin for quick profits. The third step is to never give money to anyone to trade or invest in Bitcoin on your behalf. To make your Bitcoin trading profitable, it is important to conduct thorough research, develop a solid trading strategy, keep a close eye on market news, diversify your portfolio, use proper risk management strategies, and keep track of your trades.

Lesson 31
Risks to watch

More people are beginning to invest and trade in Bitcoin, but with this comes several risk factors that investors and traders should consider. As a beginner in the world of Bitcoin, it is important to understand these risks. This knowledge will enable you to make informed decisions and avoid losing your money.

Some of these risks are outlined below:

Volatility

One of the biggest risks associated with investing and trading Bitcoin is its high level of volatility. Volatility refers to the rate at which the price of an asset changes over time. Bitcoin is known for its rapid and unpredictable price swings. This means that the value of Bitcoin can go up or down quickly and without warning, making it hard to predict which way it will go. This makes investing in Bitcoin risky, as investors can lose a lot of money if they invest at the wrong time or hold on to their investment for too long.

For instance, in 2021, the value of Bitcoin fell by over 60% from its all-time high, which happened in April. This caused a lot of investors who had bought Bitcoin at the peak to lose a significant amount of money. This is why investing in Bitcoin is considered a high-risk investment.

While Bitcoin can be a potentially profitable investment, investors must be prepared to handle the volatility and potential losses that come with it.

Lack of proper Regulatory Oversight

When it comes to investing and trading in Bitcoin, there's a significant risk associated with the lack of proper regulatory oversight. Since Bitcoin is a decentralized currency, it isn't regulated or backed by any government or financial institution. This means that investors and traders don't have the same level of protection as they do with traditional investments like stocks or bonds. Although this lack of regulation may seem like an advantage since it allows for more freedom and privacy, it also means that users are more vulnerable to scams and other types of losses.

For example, if someone loses their Bitcoin due to poor management of their Bitcoin wallet, they have no recourse or way to recover their lost funds. Additionally, hackers may target Bitcoin wallets or the platforms where people store their private keys (which are needed to access and spend their Bitcoin), leading to security breaches and potential loss of funds. It's essential to understand the importance of proper security measures to protect your investments.

Cybersecurity Risks

One of the reasons Bitcoin has become so popular is because its protocol is extremely secure and almost impossible to hack. However, hackers often target Bitcoin users and there have been instances where people have lost their Bitcoin to fraudsters. Therefore, It is important to understand that Bitcoin operates in a digital space and that makes it vulnerable to cyber-attacks.

To ensure your Bitcoin is safe, it is important to take some precautions. First and foremost, invest in a reliable and secure Bitcoin wallet. A Bitcoin wallet is a digital wallet that stores your Bitcoin. It is important to choose a wallet that is reputable and has good reviews. It is also important to keep your private keys safe and secure. Private keys are used to access your Bitcoin and should never be shared with anyone. Additionally, you should be cautious of phishing scams and avoid clicking on any suspicious links. By taking these steps, you can help minimize your risk of becoming a victim of cybercrime.

Limited Acceptance

Bitcoin is a good medium of exchange. However, not all businesses and merchants accept it as a form of payment. This means that even though people may own Bitcoin and want to use it to buy things, they may not be able to because the seller doesn't accept it. This can make Bitcoin less useful for everyday transactions like buying groceries or paying for a movie ticket. Additionally, this limited acceptance can also make it difficult for people to make money from their Bitcoin investments. If not many people are willing to use Bitcoin to buy things, the demand for it may not be as high, which can lower the value of Bitcoin and potentially reduce profits for investors and traders.

On the other hand, it's important to note that there are still many businesses and individuals who do accept Bitcoin, and its acceptance is growing. Some people even believe that in the future, Bitcoin and other digital currencies may become more widely accepted and used as a form of payment. For now, however, Bitcoin's limited acceptance can be a barrier for people who want to use it as a currency or invest in it.

Technical Complexity

Bitcoin is a type of digital currency that was created using a new and complex technology. It can be difficult for some people to understand how it works because it is different from traditional forms of money. For example, the technology behind Bitcoin wallets can be confusing, and many people have lost their Bitcoin due to technical errors caused by a lack of knowledge about how wallets work. It is important to educate yourself about Bitcoin and its technology before investing or trading to avoid any such mistakes.

Learning about the technical aspects of Bitcoin, such as how wallets work and how transactions are processed, can help you avoid losing your Bitcoin or even paying more than you need to in transaction fees.

Scalability Issues

Bitcoin is a digital currency that allows people to make transactions without involving banks or other intermediaries. However, one of the challenges it faces is scalability. This means that the current infrastructure of Bitcoin is not capable of handling a large number of transactions efficiently. When there are too many transactions, the processing time becomes slow, and the fees charged for transactions increase. This can make Bitcoin less attractive as a form of currency for everyday transactions since people want to be able to buy things quickly and cheaply. Therefore, it's important to keep an eye on these scalability issues when investing or trading in Bitcoin.

As more people adopt Bitcoin, the scalability issue could become more severe. This is because the more people use Bitcoin, the more transactions there will be, and the more pressure it will put on the infrastructure. If the infrastructure cannot keep up with the demand, it could lead to delays in processing transactions and higher fees, making Bitcoin less appealing as a currency. Therefore, it's important to understand that while Bitcoin has many benefits, it also faces some challenges, and you should carefully consider these before investing or trading in Bitcoin.

Market Manipulation

Market manipulation is a sneaky way that some investors or traders with lots of Bitcoin can control the market and make it work in their favor. They do this by buying or selling large amounts of Bitcoin at strategic times, causing the price of Bitcoin to go up or down. When the price of Bitcoin goes up, they sell their Bitcoin at a higher price, and when the price goes down, they buy more Bitcoin at a lower price. This creates a false impression of supply and demand, and other investors and traders are forced to follow their lead, creating a chain reaction that affects the entire market. Unfortunately, this can leave smaller traders at a disadvantage because they don't have the same resources or buying power to compete with the larger players. This can result in smaller traders losing their small holdings in the volatile Bitcoin market.

When investing in Bitcoin, it is important to keep in mind that the market can be manipulated by larger investors, and you should not be swayed by sudden price changes. Instead, focus on the long-term potential of the cryptocurrency and invest wisely based on your own research and analysis. Additionally, it is important to keep up with the latest news and trends in the market to stay informed about any potential risks or opportunities.

Legal and Tax Issues

Bitcoin is a digital currency that operates independently of governments and traditional financial institutions. This can create challenges when it comes to legal and tax issues. Depending on where you live, the status of Bitcoin may differ. For example, in some countries like El Salvador and the Central African Republic, Bitcoin has been made legal tender, meaning it can be used just like traditional money. However, in other countries, governments are still figuring out how to regulate Bitcoin.

If you are thinking about investing in or trading Bitcoin, it is crucial to be aware of the legal and tax status of Bitcoin in your country. This is because there may be existing laws that you need to follow, and failing to do so could result in legal trouble or financial penalties. To avoid these issues, it is important to research the regulations surrounding Bitcoin in your country and seek advice from a legal or financial professional if you are unsure about anything. By staying informed and following the rules, you can safely participate in the Bitcoin market.

Low Liquidity

Liquidity refers to the ability to buy or sell an asset quickly and easily, without affecting the price of the asset. When an asset is illiquid, it means that there are not many buyers or sellers in the market, making it harder to buy or sell the asset without affecting its price.

In the case of Bitcoin, the market for buying and selling Bitcoin is relatively small compared to traditional financial markets, such as the stock market. This means that when large numbers of people try to buy or sell Bitcoin at the same time, it can cause significant price swings. Additionally, because Bitcoin is a decentralized currency, there is no central authority that can step in to help stabilize the market during times of high volatility.

Thus, investing in Bitcoin can be risky because it may be difficult to buy or sell Bitcoin quickly, especially during times of high demand or volatility. This can make it hard to manage your investments effectively and could result in significant losses if you are unable to sell your Bitcoin when you need to. It is important to keep this risk in mind when deciding whether to invest in Bitcoin and to carefully consider your options before making any investment decisions

Summary

Bitcoin investment and trading comes with several risks that investors and traders should be aware of. These risks include high volatility, lack of regulation, cybersecurity threats, limited acceptance, technical complexity, scalability issues, market manipulation, and legal and tax issues. It is important for beginners to understand these risks in order to make informed decisions and avoid losing money. It is crucial to invest carefully and cautiously, educate oneself about Bitcoin technology, and be aware of the legal and tax status of Bitcoin in your country.

Course Endnote

Congratulations on completing the Hundred Million Dollar Bitcoin for Beginners course! By now, you should have a solid understanding of the basics of Bitcoin, including its technology, security measures, monetary value, and how to safely store and trade Bitcoin.

Learning about Bitcoin is important because it is a digital currency that is becoming increasingly relevant in our digital future. Understanding its potential to disrupt traditional financial systems and bring financial freedom to people who may not have had access to it before is important.

Throughout the course, you have learned about Bitcoin's key features, how it works, and its use cases. You have also explored the technology behind Bitcoin, including the blockchain, mining, nodes, and how transactions work. Furthermore, you have learned about Bitcoin's security measures, such as the 51% attack, double spending, and privacy concerns. You have also discovered Bitcoin's monetary value, including its implications as a store of value and its scarcity. Finally, you have learned how to safely buy, store, and trade Bitcoin, as well as the risks associated with doing so.

Remember that Bitcoin is a new and evolving technology, so it is important to keep learning and stay up-to-date on new developments. Keep exploring and stay curious about Bitcoin and other digital currencies that may emerge in the future.

Thank you for joining me on this exciting journey into the world of Bitcoin!